THE NEUROPSYCHOLOGICAL ANALYSIS OF PROBLEM SOLVING

Paul M. Deutsch Press

Classic Soviet Psychology Series

THE NEUROPSYCHOLOGICAL ANALYSIS OF PROBLEM SOLVING
A.R. Luria and L.S. Tsvetkova

ROBOTICS, CYBERNETICS, AND THE PSYCHOLOGY OF MOVEMENT
Three Volumes
R. Bernstein

EDUCATIONAL PSYCHOLOGY
L. Vygotsky

SOVIET PSYCHOLOGY
Two Volumes
R. Rubenstein

VYGOTSKY AND HIS POSITION ON PSYCHOLOGICAL SCIENCES
M. Yaroshevski

THE PSYCHOLOGY OF PLAY
L. Vygotsky

SOVIET NEUROPSYCHOLOGY: HISTORICAL AND CURRENT PERSPECTIVES
Video Series

The first video in this series will be a personal tribute to A.R. Luria and will feature highlights from his life and work, edited by his daughter, Elena Luria. Other videos in the series include LURIA'S NEUROPSYCHOLOGICAL INVESTIGATIONS, edited by Anne-Lise Christensen.

A continuation order plan is available for all PMD Press series, which will bring delivery of each new volume immediately upon publications. Volumes are billed only upon actual shipment. For more information please contact the publisher.

THE NEUROPSYCHOLOGICAL ANALYSIS OF PROBLEM SOLVING

By Aleksandr R. Luria
and Lubov S. Tsvetkova
Edited by Robert J. Sbordone
Translated by Aleksandr Mikheyev
and Sergei Mikheyev

Paul M. Deutsch Press, Inc.
Orlando • Helsinki • Moscow

This volume is published under an agreement with the Copyright Agency of the USSR (VAAP)

ISBN: 1-878205-10-2
Library of Congress Catalog Card Number: 90-83131

Cover designed by Joe Groom Design
Printed in the United States of America

About the Authors

Aleksandr R. Luria

Professor Luria's preeminence as one of the world's foremost contributors to our understanding of the human brain and to the eventual merging of neurology and psychology, is now firmly established. Professor Luria was Professor of Psychology at the University of Moscow and a member of the Academy of Medical Sciences of the USSR. He is author of numerous books on psychology and neuropsychology and these works are gradually becoming available to the English reader.

Lubov S. Tsvetkova

Professor Tsvetkova was a student of A.R. Luria and worked with him for 17 years. This book is a product of their collaboration, and is the last book that Professor Luria worked on before his death. Professor Tsvetkova is a Professor of Psychology at the University of Moscow, and Head of the Laboratory of Neuropsychology and Neuropsychological Rehabilitation, of the Department of Psychology at the University of Moscow. Paul M. Deutsch Press will be introducing a number of her books to the English reader for the first time in 1990-91.

About the Editor

Robert J. Sbordone, PhD, A.B.C.N.

Dr. Sbordone is internationally recognized for his expertise in clinical neuropsychology, head trauma, post-concussion syndromes, and cognitive rehabilitation. He has been elected a fellow of the National Academy of Neuropsychology and has written numerous books and publications. Dr. Sbordone has also developed several widely-used neuropsychological tests, and is one of the pioneers in computer-assisted cognitive rehabilitation. Dr. Sbordone is also Clinical Assistant Professor in the Department of Neurosurgery and the Department of Physical Medicine and Rehabilitation at the University of California at Irvine Medical Center.

THE NEUROPSYCHOLOGICAL ANALYSIS OF PROBLEM SOLVING

Table of Contents

TO THE AMERICAN READER

The book "Neuropsychological Analysis of Problem Solving" now offered to American readers was written by Professor Luria and me back in the 1960s. This analysis, however, has retained its importance given the development of neuropsychology, psychology and academic knowledge in general.

Many researchers from different scientific disciplines have shown an interest in the problem of man's intellectual activity, particularly, the topic of thinking. They have been concerned about a number of issues associated with intellectual activity. But the top priority has been given to the interconnection between the psyche, including mental processes, and the brain. For example, A. Leontiev, a distinguished Soviet psychologist, wrote in one of his books: "It is hardly necessary to convince psychologists that the psyche is the function of the brain... "[1]

The neuropsychological method can help to answer this question if we assume that neuropsychology is:

1) an *applied branch* of psychology;
2) a valid and *very sensitive method of psychology* capable of contributing to the solution of its fundamental problems;
3) a branch of knowledge proper, which has its own subject matter, and subject of inquiry which sits at the junction of a variety of divergent clinical and scientific disciplines, such as psychology and physiology; neurology and neurosurgery; linguistics and psycho-linguistics. Therefore, neuropsychology deals with a number of its own problems as well as the problems that lie at the junction of these disciplines.

This book is not limited to the study of such neuropsychological problems as the pathogenesis and the localization of an injury causing an impairment of intellectual activity; syndromes and mechanisms of impairment; and brain processes underlying intellectual activity, etc.

This book examines general psychological problems such as the structure of intellectual activity; the interrelation between different types of mental activity and their interaction; the principles of their integration, and the interconnection between the psyche and the brain.

In my view, this issue has always attracted worldwide attention, but today has come to the forefront of psychological research due to a number of innovative concepts which have been developed in Soviet psychology and primarily the teaching on activity elaborated by the distinguished Soviet psychologists L.S. Vygotsky and A.N. Leontiev.

In the context of their theory, mental processes are now treated as an activity underlying any psychic process and formation. The second reason for this neuropsychological boom lies in the rapid development of technology and, above all, computer technology. Artificial intelligence requires not only knowledge of engineering but also knowledge of the humanities and natural sciences, in particular, of psychology, neuropsychology, physiology, etc.

The correct approach to a wide range of computer problems is impracticable without proper information on the functioning of the brain and the interconnection between mental processes, including intellectual activity, and the brain. Therefore, we have witnessed the development of an interesting although heatedly-discussed scientific discipline studying artificial intelligence. The representatives of this scientific school have displayed great interest in the findings of neuropsychology.

Thirdly, concern for the problem of intellectual activity is also accounted for by the resumed discussion on the problem of the biological and the social aspects of man.

And, finally, the neuropsychological boom is also due to the increasing number of patients with intellectual disorders after local brain lesions of different etiology. A more profound study of intellectual impairments, their structure and underlying mechanisms is necessary not only for the continuous improvement of theoretical concepts but, above all, for improving our ability to help these patients. Thus, our research findings should be used to identify local brain lesions and develop rehabilitation techniques for patients with disturbed intellectual activity.

Soviet psychology regards intellectual activity as the basic form of man's cognition of his environment. Soviet psychology, on the one hand, does not view intellectual activity either as an abstraction or actions with abstract notions. On the other, it does not limit intellectual activity to biological processes. The Soviet psychological school defines intellectual activity as a complex psychological process which:

- has its own structure, a very complex cerebral organization,
- involves the most complex forms of interaction with other psychic formations, and
- relies on and interacts with the socio-historical evolution of society and man's psyche in general.

Intellectual activity occurs during the process of purposeful activity, in the course of work. It relies not only on our ability to deal with abstract symbols and relations, but also on specific thoughts. Practical intellectual abilities are just as valuable as academic powers. Intellectual activity and, above all, thinking, is essentially an ability to reflect abstract and social relationships with the surrounding objective world and phenomena. This accounts for man's ability to understand reality and creatively harness it to achieve his needs and goals. In this book Professor Luria and I proceeded from this perspective and examined the causes of its impairments within a neuropsychological perspective.

This book examines the relationship between intellectual activity and brain lesions. Even today we know very little about the most essential aspect of intellectual disorders — their cerebral organization. We have examined this complex process by studying the problem solving process which serves as a model of intellectual activity. Examining the relationship between problem solving impairments, and brain lesions allowed us to gain insight into the cerebral organization of intellectual activity.

By examining the emotional-volitional sphere of activity, particularly the motivation and purposefulness of behavior during the process of problem solving, we took another step toward understanding the psychological essence of intellectual activity.

Generally speaking, this book touches upon *two groups of problems.* Those making up **the first group** concern the neuropsychological assessment of intellectual activity as related to a number of specific issues:

- localization of intellectual activity in the brain;
- the role of different brain areas in the realization of intellectual activity, in particular, that of the frontal and parietal sections of the left hemisphere;
- identification and analysis of the factors underlying intellectual disorders;
- connections between various factors and the observed mental syndromes.

Research in these areas has brought us closer to answering the question of the cerebral organization of intellectual activity.

The second group of problems concerns the study and description of the neuropsychological variants of the frontal syndrome. We were mostly interested in:

1) the mechanisms and structure of problem solving disorders as a consequence of lesions in prefrontal, posterior-frontal, and medial-basal areas of the frontal lobes.
2) the distinctive features of problem solving disorders caused by an injury to the frontal lobe as a function of laterality, and other cortical regions.

The study of the two above groups of problems enabled us to clarify the role and contribution of the frontal lobes and parietal areas to the cerebral organization of intellectual activity as well as permit us to develop methods to qualitatively assess its influence on the patient's behavior.

The third group of problems concerns research on the *psychological aspects* of problem solving disorders and their treatment, particularly the study of the psychological structure of the disorder, identification of specific impairments within the structure of intellectual activity, and the dependence of this impairment on the localization of a brain lesion.

These problems were also examined to answer a number of academic and practical questions, particularly regarding the development of rehabilitation techniques for the elimination of intellectual impairments and formulating reliable neuropsychological methods for identifying intellectual disorders caused by focal brain lesions.

I would also like to draw your attention to the part of the book which describes the use of rehabilitation training (and its methods) for a qualitative differentiation between various syndromes and mechanisms underlying intellectual disorders, and to show their dependence on the localization of a lesion. On the other hand, we tried to show the relationship of rehabilitation training to such factors as specific neurobehavioral syndromes, and cerebral localization. We also attempted to develop rehabilitation treatment of patients with local brain lesions.

Unfortunately we could not touch upon a number of important issues in this book which were prompted by our research findings on intellectual disorders caused by brain lesions. These issues have been discussed in later publications,after the death of Professor Luria. In particular, I studied how patients *understand* a problem and its *question* and how they can formulate *the final and intermediate questions* in the course of problem solving. My research findings brought me to the study of *understanding* in general and its role in man's intellectual activity as well as impaired understanding of a problem as a possible reason for a problem solving disorder.

I interpret understanding as the product of the patient's and experimenter's (or rehabilitation specialist's) joint work in the process of problem solving as well as the result of mental transformation of the objective essence of an object (in this particular case — different problems).

With this in mind I worked on the issue of verbal problems, in particular, understanding literary texts differing in form and structure and on the issue of non-verbal — constructive problems. Studying acalculia I also worked on the impairment of the concept of number and arithmetic operations, etc.

I have described these and many other findings on further neuropsychological studies of intellectual disorders caused by local brain lesions in a series of articles and monographs.

Finally I would like to emphasize that I was honored to be Professor Luria's student and collaborator. My work with Professor Luria (from 1958 to 1975) is described in this book and many other joint publications and has not only molded my scientific character but also has had a important impact on my general world outlook.

In my previous publications I had an opportunity to express my heart-felt gratitude to Dr. Paul M. Deutsch for his interest in my work and the opportunity to acquaint American readers with my research findings.

Taking this opportunity I would like to once again thank Dr. Paul M. Deutsch for the publication of my books and this book in particular. I would also like to thank him for his noble efforts to popularize knowledge which is so badly needed by our patients and their relatives who are looking up to us, rehabilitation specialists, in the hope that we shall be able to develop ever more effective rehabilitation methods.

Professor Lubov Tsvetkova
18.01.1990

---

[1]A.N. Leontiev, Activity, Consciousness, Personality, Moscow, Political Literature Publishers, 1975, p.6 (in Russian).

FOREWORD

This book is virtually an attempt to use neuropsychological methods to analyze the process of problem solving. It goes without saying that such an attempt requires some comments.

The process of problem solving has a complex psychological structure. We first analyze a statement which formulates the objective of an arithmetic problem, then distinguish significant relationships mentioned in it and develop a certain solution scheme; thereafter we determine operations required to carry this scheme through and, finally, compare the obtained result with the original statement. The desired effect may be attained only through continuous control over operations performed.

The actualization of this complex process may be impaired in various elements. Teachers are well aware of the fact that most difficulties in problem solving faced by their pupils are caused by:

- the insufficiently careful and consistent analysis of a statement;
- the uncontrolled development of inadequate hypotheses;
- the incorrect use of stereotyped solution methods which often replace a true search for the proper program.

Their mistakes are also often caused by insufficient attention to the comparison of the solution with the original statement. Very few mistakes result from difficulties in calculations proper.

In different cases and at various training stages successful problem solving may be hindered by defects in different elements of the process described. Therefore, differing methods and techniques are needed to eliminate these defects.

Experience shows that the process of problem solving is psychologically so complicated that standard methods established in pedagogy and psychology are insufficient to precisely determine individual factors hindering effective problem solving.

Speaking on the importance of research on pathological changes in activity (in particular, the analysis of behavior changes in lesions of certain cortical areas) Ivan Pavlov notes that "pathology often reveals to us in a simplified and disintegrated form, integrated and complicated phenomena which are hidden from us in a physiological norm." According to him, we begin to study many facts "only after they have graphically manifested themselves in a pathology."[1]

This thesis is of paramount importance for the neuropsychological method of research on complex mental processes.

The study of local brain lesions, particularly within the past 20 years,[2] has shown that certain cortical areas play a different role in the organization of complex forms of mental activity. Therefore, they provide different factors which influence this activity. Hence, although an injury to virtually any cortical area may impair complex mental processes the *character* of this impairment will be different each time. A focal brain lesion disturbs a certain factor in the organization of mental processes and, thereby, impairing their normal functions. Hence, research on impairments in the complex forms of mental activity caused by the withdrawal of certain "factors" is very important for analyzing their psychophysiological structure.

For instance, it is common knowledge that the parietal-occipital areas of the brain are the central end of the visual-spatial analyzer and that lesions in these areas drastically impair the synthesis of separate irritations into *simultaneous (spatial) groups* but do not disturb the purposeful nature of mental activity.

It is also known that the temporal and frontal-temporal cortical areas are the central end of the acoustic and visual analyzers and their lesions lead to a measurable difficulty in synthesizing separate sound and motor impulses into *consecutive series,* although they do not affect the power for simultaneous spatial syntheses.

Finally, it is known that the frontal lobes which are, of course, highly developed in man, play an important role in regulating complex forms of mental activity, as they provide the retention of original intentions, inhibit inappropriate behavior and impulses and subordinate one's behavior to complex programs.[3]

Each cortical area plays a specific role in providing different aspects of complex mental activity. Therefore, it is important to study mental disorders caused by an injury to specific brain areas. By observing the *specific manifestations* of impairments in complex mental activity caused by different brain lesions we may learn *which role is played by certain factors.* This, in turn, allows us to discover through disintegration and simplification, integrated and complicated phenomena which are hidden from us in a physiological norm.

Thereby, the review of alterations in the structure of mental activity caused by focal brain lesions, which is virtually the substance of neuropsychology, becomes a *method for analyzing complex forms of mental activity.*

One of the authors has already applied this method to analyze the psychophysiological structure of writing.[4]

In the present book we attempt to use this method to analyze a much more complicated process of problem solving. We hope that research on how the process of problem solving is impaired by differently localized brain lesions, which involves different factors of complex mental activity, will help us examine the difficulties in the actualization of intellectual processes we have hitherto failed to analyze.

We have deliberately limited this study to the review of intellectual disorders caused by lesions of posterior (parietal-occipital) and anterior (frontal) brain areas as we have every ground to believe that an ability to synthesize separate elements into simultaneous (spatial) groups, subordinate one's behavior to an original plan and provide an adequate orientation basis for an activity are the most important conditions for successful problem solving. Therefore, the authors have assumed that a careful analysis of the impairments caused by the above lesions provides the most valuable information on the various difficulties which may occur in the intellectual processes under review.

We are also emphasizing in greater detail the various forms of problem solving disorders which occur in the frontal syndrome, that is, caused by differently localized and differing size lesions of the frontal lobes which lead to various impairments of meaningful and purposeful activity. These lesions, which have long been inaccessible to scientists, have been analyzed as a result of extensive research on the functions of the frontal lobes in our laboratory during the past 20 years.

The authors have not restricted themselves to a careful investigation of impairments in problem solving in differently localized brain lesions, but have also attempted to determine *some techniques for rehabilitation training* which could compensate for the impairments observed in each particular case.

It turned out that such techniques varied for differently localized foci and differing forms of impairments in mental activity. Therefore, we developed *different programs* which could be helpful for partially restoring the process of problem solving impaired by focal brain lesions.

Such an approach transformed rehabilitation training into a method for detailed research on the structure of the disorders under review. Furthermore, it enabled us to evaluate the rehabilitation principles and methods which were utilized to eliminate various problem solving difficulties caused by different focal brain lesions.

The authors view this book as one of the first and therefore, inevitably imperfect attempts to analyze in psychological terms the numerous difficulties which may occur during the complex process of problem solving. However, one should not assume that all the impairments described for patients with local brain lesions are equally characteristic of the difficulties faced by normal schoolchildren in problem solving. The neuropsychological method of research on the processes of problem solving may be successfully applied only through certain analytical work which should draw psychologists' and teachers' attention to the psychological conditions which are part of complex mental activity. The authors are convinced that a careful assessment of the facts described will allow for a thorough study of difficulties faced by pupils in problem solving. We believe that it will be an additional impetus to new studies which are so badly needed by our pedagogical sciences.

The authors would like to thank the staff of the Institute of Neurosurgery named after academician Burdenko whose facilities they used for conducting this study and Professor A. Markushevich for his valuable advice.

---

[1]Pavlov, I.P., Lectures on the Functioning of the Bigger Hemispheres, Collected Works, v.4, 1951, p.317 (in Russian).

[2]The results of other studies have been summed up in books by Luria "Higher Cortical Functions in Man" and "Human Brain and Mental Processes."

[3]This phenomenon is specifically studied in a series of articles published in a special volume "Frontal Lobes and the Regulation of Mental Processes" put out by the publishing house of Moscow University in 1966.

[4]See: Luria, A.R. Essays on Psychophysiology of Writing, Moscow, 1950 (in Russia).

Editor's Note to the American Reader

Serving as clinical editor involved the task of grappling with often circular and sometimes convoluted passages, which reflect the difficulty of translating Russian into English, while retaining the semantic integrity of the written material. As a consequence, many sentences and passages had to be rewritten because of their semantic confusion. In doing so, I tried to bear true allegiance to Luria's thinking and concepts, while also trying to use terminology which would be more acceptable and less semantically confusing to the American reader. For example, the word "impairment" was used in place of terms such as "defect" or "disorder," when the context of the sentence favored the former. Similarly, the term "experimenter" was changed to "investigator" since it appeared to be more appropriate and created less semantic confusion. In many instances, English translations of Russian ideogrammatical expressions (which often made little sense) had to be reworked, based on the editor's familiarity with Luria's previous works and past collaboration with his students. I accept full responsibility for any distortion of Luria's thinking which may have inadvertantly occurred during this process.

Luria's thinking and conceptual style reflects his genius and rich understanding of brain-behavior relationships. In this book, Luria demonstrates remarkable insight with his complex analysis of the problem solving process in patients with focal brain lesions. Luria's qualitative analysis of his patient's impairments is simply breathtaking and illustrates the advantages of this approach in formulating effective rehabilitation strategies and treatment programs for brain injured patients over traditional rehabilitation approaches, which often rely on quantitative neuropsychological test data. One then begins to realize why the latter approaches have often been ineffective with certain types of patients since the neuropsychological test data, most likely, did not lend itself to a thorough understanding of the patient's impairments and compensatory behaviors. Although this text was written 25 years ago, Luria's thinking still remains current and his rigorous efforts to both assess and treat brain damaged patients should serve as an inspiration to clinicians. Despite his enormous academic accomplishments, Luria always remained a clinician and a clinical investigator par excellence. He did not rely on technicians to examine his patients nor test scores, but rather drew from his experience of examining over 65,000 patients during his career, frequently on a longitudinal basis as the clinical course of their condition changed so that he could observe changes in their neuropsychological functioning and behavior, which he then correlated to the patients' brain-behavior relationships.

Robert J. Sbordone, PhD
Irvine, California
May, 1990

Chapter 1

THE PROBLEM AND CLINICAL METHODOLOGY

The Problem

We have every reason to define intellectual activity as organized problem solving which relies on a logical program of interconnected operations. Intellectual activity, which has, of course, been thoroughly studied by psychology, has its own peculiar structure and consists of a number of clearly distinguishable phases.

It is generally determined by some *objective*, or *the question* of the problem, which cannot be answered directly.

This objective determines the subject's entire intellectual activity, making it selective in character. An objective, formulated in the problem, is generally set on a background of a certain statement and intellectual activity begins with the *orientation to this* statement. The orientation covers:

- the assessment of the obtained information;
- the distinguishing of significant (both familiar and still unfamiliar) data;
- their comparison.

This preliminary work on the problem statement, which is the orientation basis for intellectual activity, results in the development of *a general scheme* (strategy) *for the solution* and the identification of *operations* (tactics) that have the highest probability of solving the problem and meeting the set objective. The general scheme, or the strategy, of problem solving relies on a certain system of auxiliary operations which are more or less in accordance with the objective algorithm of problem solving. It underlies individual operations which are always *selective in character* and result in a definite answer. *The verification of the answer*, or *the comparison of the obtained results with the original problem statement*, is an obligatory essential element of intellectual activity. If such a comparison testifies to the agreement between the performed operations, the obtained results, and the original problem statement, the activity discontinues; whereas if it reveals a lack of correspondence (or disagreement) between the two mentioned components the activity starts anew until the proper answer corresponding to the original statement is found.

Such a structure of intellectual activity completely reproduces the scheme which, according to modern neurophysiological data (P.K. Anokhin, 1955; N.A. Bernstein, 1957; etc.), underlies any activity controlled by the central nervous system.

The above structure of intellectual activity may reveal itself both in the operations of resolving practical (for instance, constructive) problems and in theoretical, verbal actions. In both cases during the initial stages of development this activity is realized through extended material (or materialized) actions which gradually get contracted and are first supported by loud (vocal) speech, then inner speech and finally take the form of contracted, quickly actualized "intellectual operations" (P. Ya. Galperin). The latter lays the basis for the highly automatized thinking whose structure it is sometimes so difficult to trace.

Objective research on the structure of intellectual processes and the identification of the factors accounting for the most complicated intellectual activity is a real challenge.

A subjective assessment of an intellectual act requires the splitting of the activity into simultaneous intellectual processes and does not yield reliable results. It is also hindered by the subconscious nature of intellectual processes proper. Therefore, the attempts to describe them in retrospect often result in reaching inadequate conclusions.

Much can be profited from the comparative study of intellectual activity of animals and especially from the analysis of intellectual processes at different stages of ontogenetic development provided they are conducted with the help of certain strictly accountable problems. Finally, valuable data can be obtained from a thorough description of the difficulties faced by people in the course of problem solving and the psychological assessment of their mistakes. While modern psychology has helped us develop insight into the psychological structure of intellectual activity, it is insufficient to determine which *factors* underly the process of intellectual activity, and often allows only for a phenomenological description of its structure.

Normally, all the components of the adult's (and even the child's) intellectual activity are so closely intertwined and indivisible that it is impossible to distinguish them. Therefore, a different approach is recommended: examining the changes in the actualization of intellectual processes in cases of brain pathology and especially in focal brain lesions. The latter condition may be very useful to us in this respect as the pathological focus inhibits the functioning of different factors contributing to the realization of higher cortical processes and produces different disturbances of intellectual activity.

The method of the *neuropsychological analysis of intellectual processes* was used by a number of researchers (Halstead, 1947; Pribram, 1959; Luria, 1947, 1962, 1963) and yielded important results. It allowed for the disintegration of mental processes which are normally indivisible and the *identification of the factors which underlie intellectual activity and are differently impaired depending on the location of the brain lesion.* This method made it possible to describe the dramatic differences in the impairments of intellectual processes in differently localized cerebral lesions. It also revealed drastic dissimilarities in the nature of intellectual activity caused by damage to its specific underlying cortical factors.

Our earlier research findings (see: A.R., Luria, 1962, 1963) show that the neuropsychological analysis enables the researcher to distinguish at least two basic forms of intellectual dysfunctions caused by local brain lesions. Each of them is essentially caused by damage to specific cortical centers which leads to different *types* of intellectual disorders.

An injury to the *posterior* ("gnostic") cortical areas does not, as a rule, disturb intellectual activity oriented toward a certain aim and determined by a certain objective. Nevertheless, it may measurably impair such *operations* as:

- the synthesis of consecutively received information into simultaneous schemes,
- the memorizing of the same,
- the use of special linguistic means for processing this information.

Such linguistic methods enable subjects to distinguish the required elements in the perceived material, to perform the operations of abstraction and generalization and successfully operate with memorized data.

A patient with such lesions (most frequently localized to the parietal-temporal-occipital area in the left hemisphere) always remembers a problem given to him and works hard on analyzing the meaning of the statement. His intellectual activity retains its purposeful and selective character. The results of the operations are always compared with the original statement. The patient sees his mistakes and tries to correct them. In these cases, we witness, first and foremost, the impairment of *the operations* by which the patient attempts to solve the problem. For example, successful realization of intellectual processes is hindered precisely by impaired tactical strategies.

The symptoms of intellectual dysfunctions in the lesions of the *frontal lobes* are completely different.

In these cases, the performance of well known operations is typically relatively intact. For example, such patients usually have a good command of language with its lexical, semantic and grammatical codes. They easily perform operations of spatial analysis and synthesis of perceived information and, in contrast to patients with parietal-temporal brain lesions, encounter no difficulty memorizing the material.

These patients suffer primarily from *a profound change in the overall structure of their mental activity.*

The assignments received by such patients neither generate stable intentions nor determine an unambiguous method for actualizing their mental processes. They do not analyze the meaning of the statement of the given problem; the most important stage of the intellectual act — the orientation basis for an operation — breaks down completely. People with such injuries do not work on distinguishing the essential elements of the perceived information or comparing the components of the statement. Neither do they develop the general scheme, the program for problem solving, proceeding from the preliminary analysis of its statement. Their attempts to solve a problem in most cases are limited to impulsive answers or directed to some fragments of the statement which have for some reason attracted their attention and have been distinguished from the problem statement. They encounter great difficulty in switching from one operation to another. Their stereotypes, once they have developed, are usually so inert that they hinder the transition to the following stages of the intellectual act, substituting stable clichés for the complex process of intellectual activity.

Finally, such subjects do not compare the obtained results with the original problem statement or determine the agreement or disagreement of the action's effect with the original intention. They do not notice or correct their mistakes either. Thus, the overall character of their intellectual activity becomes uncontrolled.

The impaired programming and regulation of intellectual activity as a result of frontal lesions has been described in previous publications (see: A.R. Luria, 1962, pp.241–249; 1963, pp.340–380, etc.). These particular impairments form the basis for a phenomenon which is well known to clinicians and is generally termed "the frontal syndrome." Its identification should be regarded as a major breakthrough in modern neurology and psychiatry.[1]

The description of the above intellectual disorders is just the first step forward in this most complicated field of neuropsychology.

The frontal lobes of the brain are its most complicated and latest formations. Their structure is extremely heterogeneous as they cover totally different components, each of which has a development history, anatomical peculiarities and functions of its own. An injury to different systems that are part of the brain's frontal lobes, undoubtedly, causes absolutely different disturbances and entails different *variants of the frontal syndrome*.

It is known that the *medial and medial-basal sections* of the frontal lobes constitute a specific and the most ancient area of the frontal cortex. In terms of their anatomical structure, they differ a lot from the convex sections of the frontal lobe. They are closely related to other areas of the limbic brain and the hypothalamic formations located underneath. According to our data, they contribute to the regulation of metabolic processes and control over affective conditions. An injury to the medial and medial-basal sections of the frontal lobe causes measurable affective disorders and, in most cases, causes symptoms of general disinhibition and defective regulation of different drives and appetites. But intellectual processes in these cases remain relatively preserved. Hence, it is extremely interesting to describe the secondary changes in intellectual activity caused by the lesion of the medial-basal areas of the frontal lobes.

The posterior-frontal (*or premotor*) area on the outer surface of the frontal lobe forms part of an absolutely different system. For example, numerous research findings (Fulton, 1935, 1943; Shkolnik-Yarros, 1945; Luria, 1947, 1962, 1963) have shown that the premotor area is part of the cortical sections of the motor analyzer. It is superimposed on the primary motor cortical sections and is most closely interconnected with the basal motor areas. The premotor area of the brain is responsible for the integration of the pyramidal and extrapyramidal motor systems providing for the continuity and smoothness of movements. Therefore, an injury to the premotor sections of the brain impairs the highest automatisms, hinders the transition from one element of a complex movement to another and impairs the performance of complex motor skills. If the lesion covers the premotor area of the left hemisphere and extends to the area of the second and the third frontal convolutions, it affects verbal processes with the ensuing distortions in the automatization of speech and verbal thinking, which was specifically described by Luria (A.R. Luria, 1963, pp.141–380). Specific impairments of complex intellectual processes in such lesions call, of course, for a special study.

The *anterior (prefrontal)* cortical areas have absolutely different features. We know much less about them than about the premotor section of the frontal lobe.

The data of comparative anatomy convincingly prove that the prefrontal sections of the cortex are its most recent formations. The dominance of the neuron structures in the upper cortical layers points to their complex function while their connections show that they should be related to the sections which contribute to the regulation of the most complex organic processes. Pribram and others call them the "intrinsic zones."

An injury to these sections of the frontal lobe does not necessarily produce sensory, motor or speech disorders. It may instead lead to the above impairments in the programming and regulation of the most complex forms of mental activity.

Clinicians are well aware of the lack of initiative, and the disorder of purposeful behavior, self-regulation and feedback that are caused precisely by injuries to these sections of the frontal lobes. Therefore, our concern to describe intellectual disturbances caused by an injury to this section of the frontal lobes and the identification of their underlying factors is self-evident.

It is worth noting that mental dysfunctions, which accompany injuries to the frontal lobes, *may be more or less conspicuous.*

There may be two extremes in cases of such lesions. One is illustrated by patients with a bilateral injury to the frontal lobes that occurs against a background of marked cerebral (toxic and hypertension-dislocation) manifestations. Severe impairment of purposeful behavior with marked akinetic-abulic syndrome and disordered self-regulation and feedback results precisely from such lesions. This picture is most apparent in cases of rapidly growing intracerebral tumors in the frontal lobes which extend to the corpus callosum.

The other extreme results from slowly growing extracerebral (and sometimes benign intracerebral) tumors in the frontal lobe which do not destroy but only compress brain matter. They are in many cases accompanied by very indistinct symptoms, but are sometimes without any symptoms.

We should certainly study these forms of intellectual disorders and attempt to distinguish both the common and the differing symptoms characteristic of these variants of the frontal syndrome.

Finally, we should touch upon the last aspect of the issue under review.

Lesions of the frontal lobes may occur both against a background of general inhibition and a torpid state and on a background of general agitation, disinhibition and euphoria. We do not yet know the reasons for the fluctuations between these two poles in pathological behavior changes. We are not even sure whether these symptoms result from changes in the physiological condition that may take place irrespective of the focal lesion or if the two alternatives are related to the localization of the lesion focus and are primarily manifested in the lesion of certain systems. Clinical evidence shows that disinhibition and euphoria are especially frequent in injuries to the medial-basal sections of the frontal lobe while they are rather infrequent with lesions of the posterior frontal cortical areas. Therefore, today we do not have enough evidence to reject the localization hypothesis altogether as a cause of this form of pathological behavior although we are still ignorant of the physiological mechanisms which cause this pathological form of behavior to occur.

It is only logical that we should carefully analyze the description of the intellectual changes peculiar to both forms of pathological behavior in patients with the frontal syndrome.

Hence, we shall only limit ourselves to the description of the changes in intellectual activity caused by injuries to the basal sections of the frontal area, the premotor zone and the prefrontal sections of the frontal lobes. To this end we shall use case examples to illustrate changes in intellectual processes with variants of the frontal syndrome.

Clinical Methodology

The methods used to study conceptual thinking (similar to those used by the Wurtzburg school) are inappropriate since they establish only the level of abstract codes mastered by the subject which are not indicative of the dynamics of intellectual processes.

The classification methods which were once successfully used in normal and abnormal psychology (see: Vygotsky, 1934; Goldstein, 1948; etc.) give very different results depending on the patient's premorbid level and do not always provide consistent data.

The methods of research which are adequate for analyzing so-called nonverbal intellectual operations cannot be applied to study the system of verbal thinking or its dynamics.

Therefore, we decided to use a very simple method which, common as it was, turned out to be adequate for getting sufficiently clear information on the structure of intellectual activity and the changes caused by impairment of its dynamics. By this we mean the *analysis of the solution of relatively elementary arithmetical problems,* which, in our view, is one of the best methods for objective research on intellectual activity.

Problem solving is the most definite and complete manifestation of intellectual activity. The structure of problem solving reproduces the features that are characteristic of any intellectual act. A thorough analysis of the process of problem solving under different conditions produces an opportunity to describe *the structural changes in this process* and distinguish *different factors* which determine the subject's intellectual processes. It is self-evident that such factors as poor retention of the original problem statement, weakened or impaired attention or concentration, difficulties in planning or retention of the plan, impaired performance of the required operations and, finally, defective comparison of the obtained results with the original problem statement will have different effects on the process of problem solving that can be determined by careful examination. Furthermore, the study of the problem solving process, which best serves our research purposes, has other significant advantages.

A study which relies on the application of this method may be conducted at a level that is equally accessible to *any normal subject,* even with a relatively low premorbid educational level whereas *the variations in the problem statement* may provide the conditions for identifying *differences in the structure of the impaired intellectual processes under study.*

Therefore, the success of the study depends not only on the use of this method and the careful analysis of the overall process of problem solving but also on *the choice of the series of problems.* Their solution algorithm should meet certain requirements and provide for consistent research into intellectual activity at different levels of complexity.

This criterion accounts for the need to use a certain *set of problems,* whose structure should gradually grow increasingly more complicated, thereby, providing an opportunity for a comprehensive study of intellectual processes. In selecting this set, we were guided by the principle of the continuous complication of the solution algorithm which entails distinct types of psychological difficulties.

(1) The most elementary group is made up of so called *simple problems* (of the $a + b = x$ or $a - b = x$ type).

For example: *Ann has five apples and Pete has four apples. How many apples do both have?* Or *A total of 47 pine and oak boards were recently purchased. There were five oak boards. How many pine boards were purchased?*

It is clear that such problems can be solved in one arithmetical operation (the complexity of such an operation may differ depending on the semantic content of the problem) and that *the statement of such problems should clearly determine the correct algorithm for their solution.*

(2) The simple problems described require an elementary solution. They may, however, become much more complicated by the use of *simple but inverted problem format* (of the $a - x = b$ or $x - a = b$ type).

For example: *A boy had 12 apples; he gave some of them away. After that he had eight apples left. How many apples did he give away?*

Or: *There were some birds on the tree. Three birds flew away; after which five birds were left. How many birds were there originally on the tree?*

While the operations required by these problems do not drastically differ from the first series of problems, their psychological structure differs considerably.

Whereas the first group of problems did not require any complex orientation since the statements themselves clearly determine the method utilized to arrive at the correct solution, the subject immediately encounters difficulty while dealing with the statements of the second group of problems. This difference explains the difficulty or *conflict* faced by the subject solving these problems. To arrive at a correct solution the subject must overcome the tendency to rely on the particular sequence of data contained in the statements. Patients with perservative tendencies are likely to have serious difficulty solving this group of problems.

Therefore, simple but inverted arithmetic problems can be successfully used for studying impaired intellectual processes.

(3) The simplest variants of so-called *composite problems* $a + (a + b) = x$ or $a + (a - b) = x$ or of a still more difficult type: $a + ab = x$ can create additional difficulty.

For example, *Ann has five apples and Kate has two apples more (or less) than Ann. How many apples do both girls have?* Or

Pete has three apples while Nick has twice as many. How many apples do they both have?

These problems share one important feature: they cannot be solved by a single arithmetic operation. To solve such problems correctly, one should first find the value of the addend. Only then is it possible to get the required sum. As a result, the problem statement *per se* does not determine the possible method for its solution while a simplified understanding of the statement [that is, the substitution of the absolute value of the addend for the relative one ("Kate has two apples" instead of "Kate has two apples more than Ann")] may significantly alter the solution algorithm by turning the problem of the $a + (a + b) = x$-type into that of a more simplified $a + b = x$-type.

It is characteristic that any impairment during the preliminary analysis of the problem statement can easily lead to such a simplification of its structure and results in an inadequate solution algorithm. Research on the initial stages of problem solving has shown that patients with this problem face serious difficulty if their tendencies to arrive at a simplified understanding of the problem statement is not inhibited.

(4) The next, very diverse group of problems is that of *complex composite problems* of the type:

$$x = a \times b; y = \_\_ : n; z = x - y$$

or

$$a + (a + b) + [(a + b) - c] = x.$$

For example, *A son is 15 years old. His father is 25 years older. His mother is five years younger than the father. How old are all the members of the family taken together?*

Or:

Some collective farmers have a plot of land 20 hectares large. They harvested three metric centners of grain per hectare. One third of the harvest was sold to the state. How many centners of grain were left to the collective farmers?

The two problems have an important feature in common, besides a much larger number of components than the elementary problems mentioned above: the algorithm of their solution separates into a measurable number of consecutive operations, each of which results from the previous one. Therefore, the subject should memorize the whole cycle of operations to be performed, retaining in his memory the result of the previous operation and turning it into the initial data for the following one.

The difference between the two problem variants above lies in the fact that the former requires a series of operations with homogeneous terms while the latter has heterogeneous ones (hectares and metric centers). The subject, solving the latter problem, should refer the obtained data to one of these denominations and, thus, involve them in the respective semantic system of the problem. The second problem variant has another significant distinction: whereas the first problem presupposes a series of similar operations (addition or subtraction), the second one requires the transition from one type of operation (multiplication) to another (division) and still another one (subtraction).

Dealing with different variants of such problems, the subject should not only find the adequate intermediate operations but also display sufficient stability of mnemic processes: he has to remember the sequencing of all the elements of the operations which form part of the problem algorithm.

It goes without saying that the understanding of individual data, which is part of the problem statement, alone does not provide the unique plan for its solution. The system of operations should be selected out of a large number of alternative options.

(5) *Complex problems with an inverted sequence of operations* border on the group of problems described above. One of the principle terms is unknown and should be found through a special series of operations. Such problems include an element in the inverted sequence of operations of the following type:

$$a + b = x; xm = y; y - b = z$$

For example: *A son is five years old. In 15 years his father will be three times older than the son. How old is the father now?*

Problems of this type include some elements of Group Two, which require operations on unknown terms and the inversion of the operation sequence, and of Group Four, which cover several auxiliary operations.

The second component of this problem (father's age in 15 years) cannot be immediately derived from the available initial data. It can be learned only by two consecutive operations: we should first learn the son's age in a certain period of time (5 years + 15 years = 20 years) and then multiply the thus obtained term by the value mentioned in the problem statement (20 x 3 = 60). The answer to the final question (father's age now) can, of course, be found by an operation which is opposite to the previously performed one (the operation of subtracting 15 years instead of the earlier performed operation of addition).

It is characteristic that the problems of this group have one special feature in common: all the intermediate operations that are not formulated in the statement are of auxiliary character and the answer to the final question is the product of a *series* of auxiliary operations.

It is, of course, extremely difficult to solve such problems unless we inhibit attempts to solve them impulsively and determine the series of auxiliary operations that are only implied in the statement. This is even more so because of the inversion in the sequencing of the operations in case of the last operation.

(6) The next group of problems differs from the previous problem by emphasizing the need *to compare two equations and distinguish a special auxiliary operation*, which provides the initial data for the correct solution of the problem. This group is represented by the problems of the $x + y = a; nx + y = b$ type.

For instance: *A pen and an ABC book cost 37 cents together; two pens and one ABC book = 49 cents. How much do a pen and an ABC book cost separately?*

A more complicated variant of such problems is that of the following type:

$$x + y + z = a; x + y = b; y + z = c$$

For example: *Three boys caught eleven kilograms of fish. The catch of the first and the second boys put together equalled seven kilograms while the catch of the second and the third boys amounted to six kilograms. How many kilograms of fish did each boy catch?*

Both problems share one special feature: all their individual components are unknown and must be derived only by comparing several equations which make up the problem. Thus, the subject, solving such problems, should from the very outset, operate not on individual values but on the basis of groups of equations. Moreover, each result is derived from different operations (direct subtraction of the group data from the total sum and the addition of the obtained results with their following subtraction from the initial sum).[2]

It is clear that the solution algorithm of such problems presupposes not only the retention of the entire system of data but also the performance of several operations which do not give a direct answer and are of special auxiliary importance.

(7) So-called *conflicting problems* are the most difficult ones.

These problems, besides the already described difficulties, are fraught with a specifically psychological difficulty that results from the conflict between the solution algorithm of the problem and an inert stereotype. The problem can only be solved correctly if this stereotype is eliminated. Structurally, conflicting problems may belong to any of the above groups but they are distinguished by one common feature: the statement of the problem creates a tendency to resolve the problem techniques that are in conflict with its algorithm.

For example: *A father is 49 years old. He is 20 years older than his son. How old are the two of them together?*

Or: *A worker received wages of 120 roubles and usually gave 70 roubles to his wife. But this time he gave his wife 10 roubles more out of his wages. How much was left?*

The words "older" and "more" urge the subject to perform the operation of *addition* while the idea of the statement prompts the opposite operation — that of *subtraction*. In both cases the problems can only be solved correctly provided the inhibition of the stereotyped operation (e.g. addition) which was assumed as a result of a literal understanding of the wording of the problem.

Subjects encounter similar difficulty in solving different type of problems: *A pencil is 15 cm long; its shadow is 45 cm longer than the pencil. How many times longer is the shadow than the pencil?*

The directly perceived numbers prompt the operation of division ($45 \div 15 = 3$). This tendency should be inhibited by a careful analysis of the statement and learning the relative value of the other term (*45 cm longer*). This operation presents even greater difficulty due to the need to disengage oneself from the images of two specific items (the pen and the shadow) and switch to an abstract value of length which in the case of the second item is of an additive nature (the length of the pen + an additional surplus).

Experience shows that in these cases, subjects encounter serious difficulty as they must overcome the tendency to perform the operations prompted by their literal understanding of the statement and develop a solution algorithm as a consequence of a preliminary restructuring of the problem.

(8) The last group on the list is represented by so-called *standard problems.*

They are distinguished by the need to use a special technique of purely auxiliary nature.

This group of problems may be exemplified by the problems of the following type:

$$x + y = a, \text{ where } x = 2y \text{ or}$$
$$x + y = a, \text{ where } x = y - 2$$

For example, *there are 18 books on two bookshelves. There are twice as many books on one shelf than the other. How many books are there on each shelf?*

Or let us take the problems on the difference: *There are 18 books on two bookshelves. But on one shelf there are two more books than on the other. How many books are there on each bookshelf?*

These problems cannot be solved directly by halving the sum (two bookshelves). Whereas the former requires the additional calculation of the number of parts by not dividing in two specific groups but by dividing the sum into three parts, the latter problem requires a preliminary subtraction of the "surplus" from the sum and later adding this amount to one of the divided equal parts.

It goes without saying that these problems require an additional operation. But the nature of these problems allow us to examine how easily a patient can grasp their requirements and inhibit attempts to solve them by the direct methods which they have learned in the past. The two problems also enable us to trace the ease with which the subject can disengage himself from their seeming similarity and distinguish the differences in the required algorithms.

In order to describe the structure of intellectual disorders in local brain lesions and identify their underlying factors, research on the changes in intellectual activity should be conducted on a systematic basis in order to obtain the fullest possible information. It goes without saying that a mere mention of the fact that the subject could (or failed to) solve the problem is inadequate for our purposes. Instead, we need a detailed analysis of the entire process.

This can be done by defining each of the stages of the process and altering the statement of the problem.

The study of the problem solving process should begin with an analysis of the patient's verbal repetition of the recalled statement. At this stage we can see if he remembers *the essential elements of the problem,* and recognizes the required operations, or takes in only some separate fragments of the statement which he fails to integrate into a meaningful analysis. It is equally important to determine if the patient can recall the question, since disorders in the retention of both problem components — the statement and the question — are just as important in analyzing the patient's intellectual processes as the retention of nominative or predicative phrase structure when analyzing speech disorders.

When examining the peculiarities of the patient's intellectual activity we must pay special attention to analyzing *his approaches to problem solving* and the structural format utilized during his *orientation to the problem.* As mentioned above, the pathology of higher cortical processes reveals different types of disturbances of this important process. On the one hand, it gives the investigator an opportunity to distinguish the various disturbances when the patient actively attempts to grasp the necessary operations implied in the statement of the problem time and attempts to solve the problem only after its structure is sufficiently clear. In these cases we may speak about certain failures in the performance of the subsequent operations but it is clear that the patient preserves their orientation to the problem. The pathology of higher cortical processes also reveals, however, the types of impairments that occur when the patient does not develop a preliminary orientation of the problem statement. In these cases, the patient tries to solve it, having only grasped some separate fragments of the statement without any careful analysis. In this case, the orientation basis for intellectual operations is disturbed. It is very important to study these impairments (which most frequently occur following lesions of frontal lobes) with utmost care.

While studying the distinctive features of intellectual processes it is important to examine how the patient formulates the *plan, or the general scheme, of problem solving,* and how the formulation of the preliminary plan is related to the further solution process.

Among patients with local brain lesions there is a distinct group of people (mostly with injuries to the parietal-occipital brain areas) who do not encounter any serious difficulty in drawing up a general scheme for problem solving. Instead, their difficulties involve the consistent *implementation* of this plan. Their impairments are caused by the disordered performance of separate operations, defects in memorizing all the elements of a problem and the impaired wording of respective operations.

In addition to this group, there is also another clearly definable group of patients who fail to develop any preliminary plan. Even when requested to draw up a respective plan such patients *immediately attempt to perform certain operations,* which neither rely on any general solution scheme nor correspond to any precisely formulated strategy. We are especially interested in such cases, which most often occur following lesions to the frontal lobes.

In this case, we can clearly distinguish at least two groups of patients with focal brain lesions with disturbed higher cortical functions. The patients of the first group generally retain their ability to systematically search for the required solution methods. The operations they apply to solve a problem never lose their selectivity or go beyond their formulated plan. Such patients, however, may encounter serious difficulties in identifying or performing the required operations but their actions are always determined by the original task and are perfectly meaningful. The patients of the other group easily disengage the operations developed in the course of problem solving from the original problem statement. Their operations are easily affected by extraneous factors and quickly lose their selectivity.

The impairment of the selective nature of operations may in turn assume different forms. In some cases the sequence of operations determined by the statement of the problem may be replaced by *the inert reproduction of an established stereotype.* Such substitution may be manifested both in the reproduction of the earlier derived numbers, denominations or operations and in the stereotyped reproduction of whole plans or strategies, earlier used by the patient in solving other problems. In other cases the loss of the selectivity of operations may be accounted for by the uncontrolled recurrence of *irrelevant associations* which divert the patient from the logical solution of the problem. In still other cases, this loss results from the disintegration of the patient's structurally complex program of operations. His operations are no longer determined by the synthetic statement of the problem, but are instead directly influenced by isolated *fragments* of a statement and, as a result, are impulsive or fragmentary in character. It will be shown that the disorders in the process of problem solving, peculiar to patients with injured frontal lobes, most often refer precisely to this category and the different forms of impairments, outlined above, may be especially prominent in different variants of the frontal syndrome.

Finally, it is extremely important to study how such patients *understand what they did and attempt to correct their mistakes.* With these operations in mind patients with impaired higher cortical processes fall into two distinct groups. Patients in the first group are perfectly aware of everything they do to solve a problem. They correct their mistakes through the systematic analysis of the applied techniques. These patients are not only able to clearly describe the techniques used to solve a problem but always retain a conscious attitude to the obtained results. As a consequence they are able to compare them with the original statement of the problem and correct the mistakes made.

The patients from the other group behave in a completely different manner. Even having performed the required operation correctly, they cannot explain why it was performed this way or which methods they used. When asked such questions, they generally simply repeat the performed operation. They substitute this repetition for the process of analyzing the performed operation as it is inaccessible to them. Such patients, as a rule, cannot compare the result obtained with the original statement of a problem and, therefore, can neither see their mistakes (they either assess the result as the correct one or altogether refuse to assess it in any way) nor correct them. It will be shown below that there is a large group of patients with marked frontal syndrome who suffer precisely from such symptoms of intellectual disorder.

Although the study of changes in intellectual activity depends on the use of a constant set of problems, characterized by growing complexity, the analysis procedure should by no means remain invariable. It must always rely on the active involvement of the investigator who must alter the statement of a problem during the course of the experiment, ask respective questions and help the subject in some way.

The systematic use of *assistance,* (e.g. cues or prompts) or the training of the patient, the assessment of how the subject uses these cues and which particular training methods enable him to compensate for his impairments allow the investigator to understand the significant features of the patient's intellectual processes. Especially interesting information on the patient's specific intellectual processes can be gained when he goes from attempts to solve a problem on his own to solving it with the investigator's help. The investigator helps the patient divide the process of problem solving into separate questions, or stages and program his behavior. The investigator's prompts to the subject of the methods of solution which the latter cannot find himself are valuable sources of data. The investigator also watches to which extent the patient is able to transfer these methods to new problems. In this case, training of the patient can be seen within a research context.

In order to show the role of training in our study, we shall summarize the psychological characteristics of training, as a method for investigating the process of solving arithmetic problems.

Research on changes in the higher mental functions in the evolution of dysfunction is one of the most valuable methods for studying the pathology of cortical functions. Used for these purposes, training may contribute to a more careful analysis of the intellectual disorders characteristic of patients with focal brain lesions.

A thorough neuropsychological assessment of their problem solving disorders revealed central defects underlying their intellectual impairments. Based on this data, we developed a training methodology to determine a basis for compensating for the patient's impairments. The training methods should be in *perfect harmony with the main features of a disorder* which vary in different local brain lesions. If the training method (peculiar to each group of patients) actually suits the disturbance, it will permit compensation of the impairment.

To achieve this, we attempted to use the principles of programmed education to ensure strict and systematic control of the restoration of mental functions.

Proceeding from the theoretical concepts of programmed education developed by Leontiev and Galperin (1964), we used a system of programming for consistent analysis and problem solving, since programmed operations should aim at restoring the defective link in intellectual activity. We first attempted to identify the operations whose consistent performance could produce the restoration of activity as a whole. Our programs consistently identify the consecutive operations necessary for effective problem solving.

It will be shown below that in this part of the study as well, the patients separated into different groups *each of which profited from different programs.*

Patients with parietal-occipital lesions needed programs that laid the basis for performing the operations of analysis and solving arithmetical problems. Those with the frontal syndrome profited more from the programs which organized their *behavior* in the situation of problem solving, provided the basis for the orientation and research activity, and helped them to control their operations.

Our research has proven the effectiveness of such programmed training. The programs used restored the proper structure and the dynamics of impaired intellectual operations. Information obtained during the experiments of programmed rehabilitation training contributed heavily both to an adequate academic analysis of intellectual disorders caused by brain lesions and science-based practices of restoring the impaired intellectual functions.

All the methods described lay the necessary basis for the structural analysis of disordered intellectual processes while the structural analysis *per se* enables the investigator to identify factors underlying different forms of intellectual disturbances caused by different focal brain lesions and, therefore, to meet the main challenge of the research: to determine the constituent factors of normal human intellectual activity.

---

[1]The functions of the brain's frontal lobes in the regulation of mental processes are analyzed in the collected papers edited by Luria and Kholmskaya "Frontal Lobes and the Regulation of Mental Processes."

[2]Here and further on we analyze one of the solution options that was generally preferred by our patients, without assessing other possible solution variants.

Chapter 2

IMPAIRMENTS OF PROBLEM SOLVING WITH LESIONS OF THE PARIETAL-OCCIPITAL AREAS

We shall begin the discussion of intellectual disorder produced by focal brain lesions with the assessment of impaired problem solving caused by injuries to the parietal-occipital areas of the left hemisphere. Thus, we shall describe a possible type of intellectual disturbance which will later be contrasted to intellectual disorders caused by injuries to the frontal lobes which are the main concern of this book.

An Analysis of the Problem Solving Process Caused by Injuries to the Parietal-Occipital Brain Areas

The parietal-occipital sections of the human brain are, of course, heavily involved in the analysis and synthesis of perceived visual information. Since the parietal-occipital brain areas coordinate visual, kinesthetic and vestibular analyzers, they play an important role in integrating the received information into simultaneous spatial groups. This is done through comparing separate signals and organizing them into aggregate, spatially-oriented structures. Hence, an injury to these cerebral areas deprives man of an ability to perceive the elements of complex spatial structures in their respective operations and organize the perceived material into a single aggregate (Head, 1926; Potzl, 1928; Pribram, 1959; Ajuriaguerra and Hecaen, 1960; Luria, 1962, 1963). This basic impairment leads to a variety of complex disturbances which are well known to clinicians.

Patients with injured parietal-occipital brain areas often fail to perceive simultaneously a complex set of figures presented to them, since their attention is usually confined to one of the perceived elements. They encounter difficulty synthesizing the elements into a single "gestalt." Their perception becomes narrowed and fragmentary in character. For example, patients with simultaneous agnosia have to move their eyes from one element to another and simply guess about the relationship between individual elements (Luria, 1959, etc). Impaired simultaneous synthesis results in the marked difficulties in visual-spatial analyses: such patients are unable to find their bearings, lose the power to clearly differentiate between right and left, make blunders in fulfilling constructive tasks, fail to read spatial coordinates on the geographical map precisely, etc. Clinicians are well aware of these disorders underlying the syndrome of spatial agnosia and apraxia which have been discussed in numerous publications (Ajuriaguerra and Hecaen, 1962; Luria, 1962, etc.).

It is also important that defective simultaneous synthesis which is the central symptom of the occipital-parietal lesions leads to impairments that go far beyond visual perception and spatially organized operations.

Sechenov noted (1878) that numerous intellectual processes can be normally realized only if "the elements of thought" are integrated into simultaneously visible "groups." Such an integration of elements into organized simultaneous groups occurs while developing concepts and during internalized reproductions of the complex visual imagery which lays the basis of the human psyche.

The significance of such internal spatial schemes for situational orientation has been repeatedly stressed in literature (P. Jeanet, 1928; F.N. Shemyakin, 1954, 1955). The possibility of harmonizing concepts with a certain "quasispatial scheme" is a basic condition for *mathematical thinking* which always operates with some topologically organized notions. It is also common knowledge that any computations and especially mental arithmetic, dependent, of course, on so-called table computation, generally use internal spatial schemes into which the number system is fitted. It goes without saying that any arithmetic operation (for instance, addition and subtraction, especially if it requires carrying over from the tens column) should without fail be performed within a certain internal system with a peculiar spatial structure in which the subtrahend is put on one side (usually to the left) and the addend on the other (usually to the right). Arithmetic operations (which can, of course, be traced to visual geometric imagery) are nearly impossible without the support of internal spatial schemes.

The internal "quasispatial schemes" also contribute to the actualization of some special forms of *verbal thinking*. As already noted in literature (Head, 1926; Luria, 1946, 1947, etc.), the idea of certain logico-grammatical structures cannot be grasped without understanding the relationships between their constituent words. If the human brain did not simultaneously synthesize the meaning of individual words into the semantic systems which are the essence of logico-grammatical constructs, a rapid, nearly immediate comprehension of their meaning would be impossible. Modern linguistics is well aware of the need to involve perceived speech into some familiar systems of logico-grammatical codes. But for the involvement of separate, consecutively perceived elements in the system of logico-grammatical codes it would be impossible to understand the meaning of logico-grammatical structures made up of separate grammatical constructions. For the same reason, a construct in which inflexions, prepositions or some other means are used to *govern* the constituent words would inevitably disintegrate into the isolated word meanings of its components. Prepositional constructions "the pencil *to the right of* the pen," "the circle under the cross," "spring *before* summer" or inflectional constructions "father's brother," "daughter's son" can be understood only due to complex work designed to divert one's attention from the direct meaning of separate words and concentrate on the *relationship* expressed by the construction since the latter bears the main idea of the perceived information.

Hence, *an injury to the cortical areas which provide the processes of synthesizing consecutive elements into simultaneous (spatial) schemes will inevitably impair the intellectual processes actualized with the direct involvement of such "internal schemes."*

It is known that patients with lesions of the parietal-occipital areas of the dominant (left) hemisphere exhibit other symptoms, besides those of disordered simultaneous visual perception and defective spatial orientation. They, as a rule, easily forget the names of the fingers (to remember them one should distinguish the interrelations between the given finger and the others). They encounter difficulty determining the positional value of numbers mistaking the tens column (the second digit to the right) for the hundreds one (the second digit to the left) in four-digit numbers. They are very bad at mental arithmetic which requires the retention of the interrelations between numbers in the short-term (operative) memory. Such patients confuse the opposing arithmetic operations of addition and subtraction, multiplication and division as well as the signs denoting these operations. The defects manifested by this group of patients are not, however, limited to the above calculation disorders. The disturbance of simultaneous (spatial) synthesis, caused by injuries to the parietal-occipital sections of the left hemisphere, inevitably results in *a severe impairment in the understanding of logico-grammatical relationships.* This impairment has been thoroughly studied by many scholars (Head, 1926; Luria, 1947, 1962, etc.) and given the name of "semantic aphasia". Patients with this syndrome easily figure out the meaning of separate words and simple grammatical constructions which do not require decoding of complex forms of grammatical rules. But such aphasics fail to understand the meaning of structures if they are made up of words that enter into complex grammatical relations with the help of prepositions, conjunctions and inflections or require abstraction from the direct meaning of their constituent words and the distinguishing of special logico-grammatical relationships.

Such people encounter serious difficulty in analyzing phrases like "the cross under the circle" versus "the circle under the cross" or "fathers' brother" versus "brother's father." They generally say that the meaning of each separate word is perfectly clear to them but they cannot figure out their interrelationships. Therefore, when asked to draw "the cross under the circle" they usually do it the way the words are used in the phrase. They first put a cross and then draw a circle under it. It is equally difficult for such patients to distinguish the difference between the phrases "father's brother" and "brother's father". The aphasics say that both of them refer to a father and a brother but confess that they cannot understand the meaning of either of them, to say nothing of discriminating between them. As a result of their inability to grasp logico-grammatical relationships immediately, such patients cannot figure out the idea of the sentence "There is a bird's nest on the branch of the fir-tree." Here is how they try to decode the meaning of the whole sentence: "On the branch . . . here . . . on the branch . . . and why of the fir-tree? . . . on the branch . . . and the fir-tree . . . and the nest, into the bargain . . . And what does the whole phrase mean?!" They are, of course, unable to comprehend a text with its more complicated logico-grammatical structure. They may spend hours on work that is normally performed in a couple of seconds and still fail to obtain the desired results.

It is characteristic that the disorder of simultaneous (spatial) synthesis, as well as the numerical and logico-grammatical operations associated with it, is manifested against a background of perfectly intact meaningful and purposeful behavior by such patients. They remember the task well, attempt to find methods for its fulfillment and continuously compare the results obtained with the original intention. Aphasics realize the arising difficulties and try to correct their mistakes.

As a result, their structurally intact activity is disturbed only when they perform *operations* that require the involvement of the impaired function. Such patients suffer from clearly distinct defects of the intellectual processes.

The above mentioned features determine the difficulties faced by this group of patients in solving arithmetic problems.

People with parietal-occipital lesions are, as a rule, very attentive both during listening to the assignment and performing the task. Having listened to the instructions and statement of a problem, they usually ask questions to clarify them and sometimes request the therapist to repeat the problem, especially the parts which they either "did not quite understand" or "forgot." Such intact initial orientation to the assignment provides all the necessary conditions for going from the analysis of the statement to problem solving proper.

However, when they come to analyze the problem and, above all, its grammatical structure patients with occipital-parietal lesions immediately exhibit impairments which prevent them from performing the necessary operations for understanding the problem and correctly solving it. Therefore, in their case, work on an arithmetic problem begins with *the detailed grammatical analysis of its statement.* Such patients try to figure out difficult logico-grammatical constructions with an eye to disclosing their logical essence and grasping the idea of the problem. As they suffer most from the misunderstanding of comparative, attributive and prepositional constructions, it is only natural that phrases like "the father is older than his son," "the shadow is longer than the pencil," "the shopkeeper got x roubles *for* the goods sold *in* one day," etc. present a serious (and often insuperable) difficulty for such patients and require special work. Their disturbed capacity for the immediate understanding of the meaning implied in certain logico-grammatical constructions causes primary difficulties faced by these patients in analyzing arithmetic problems despite their perfectly intact stable purposeful attention.

Our research has proven that once these patients manage to figure out the meaning of the grammatical structure of the problem statement, either on their own or with the investigator's assistance, they immediately get down to specific work on its resolution. The patients distinguish the question in the problem and its statement. They compare its components with each other and with the final question of the problem, distinguish the available data and formulate a logical solution plan. Such an analysis, as a rule, enables such aphasics to develop a general solution scheme. Many of them are able to determine the number of operations needed to solve a problem even before they have solved it in writing. But the moment such patients get down to specific operations, they face additional difficulties in translating a logical solution plan into mathematical relationships and finding adequate arithmetic operations.

When aphasics begin resolving a problem by performing the respective arithmetical operations, they encounter difficulty in understanding some grammatical constructions and translating them into mathematical terms. Such patients often face difficulty in determining arithmetical operations which would correspond to the logical interrelationship of the entities mentioned in the problem even when it might appear as if there were no complex logico-grammatical constructions in the text of the statement.

Such are the central defects of intellectual disorders in parietal-occipital lesions.

If our patients managed to identify the required arithmetic operations (either on their own or with the therapist's help), having decoded the grammar of the problem, they solved it correctly. Moreover, throughout the lengthy search period they showed fully-preserved, stable attention, selectivity of operations, and the controlling function of intellectual activity.

Let us briefly summarize those aspects of logico-grammatical constructions used in arithmetic problems which may present the greatest difficulty for this group of patients.

There is a long-established tradition of formulating arithmetic problems. The statement generally has a contracted logico-grammatical structure and contains some turns of speech which are directly related to specific arithmetic operations. For example, our memory always associates prepositions "by" and "into" with the operations of multiplication and division, respectively, while the preposition "by" may be also associated with the operations of addition and subtraction. The words "two-fold," "three-fold" and "equally" trigger off the operations of multiplication or division while the affirmative word combination "so many more (less)" requires the operations of addition (subtraction). Detailed questions beginning with the words "by how many times" are associated exclusively with the operation of multiplication while the questions to the problems beginning with the words "how much less" are unambiguously related to subtraction.

These absolute associations between some words and certain arithmetic operations exist in the background of every educated person. They do not, however, determine the method of problem solving. They simply prompt to us the range of operations that may be used and provide certain alternatives in the search for the proper operation.

The same prepositions, words and sometimes whole word combinations usually require different arithmetic operations depending on the *context.*

Hence, complete information is provided not by separate words and constructions but by *whole logico-grammatical structures* which make up the coherent text of a problem. Therefore, to solve a problem correctly we should first find the specific meaning of a polysemantic word and only then determine which arithmetic operations correspond to this specific meaning.

Let us consider the two following problems:

1) A collective farm sent five trucks of eggs to the city — 720 eggs in each. How many eggs did they send to the city?
2) All in all the collective farm sent 120 trucks of eggs to the city, sending five trucks each day. How many days did it take the collective farm to deliver all the eggs to the city?

In both problems we come across the grammatical construction "so many units each." But whereas in the first problem it requires the operation of multiplication (720×5), in the second one it requires division ($120 \div 5$).

Or, let us analyze the statement of the following problem: First time mother bought two kg of pears, the second time she bought two-times more pears than the first time and twelve kg of apples. How many times more apples than pears did she buy the second time?

The word combination "n-times more" used in different contexts requires opposing operations: multiplication in the text of the statement and division in the question.

A problem with the inverted type of solution asks for even more complicated activity: "There were some pears on the plate. After 7 pears were added to them there were 15 pears on the plate. How many pears were initially there?"

The thing is, that the verb "to add" triggers off the image of an increase. Therefore, it would be logical to assume that it directly prompts the operation of addition. We should, however, take into account the implied meaning of the problem which in the given context actually requires its opposite — the operation of subtraction.

Hence, alongside problems which mandate the use of intent stereotypes, based on one's previous experience, there are problems of a different type that require more complicated work on decoding their grammatical structure. In dealing with such problems it is necessary to gain an insight into their implied meaning and be able to abstract from the direct meaning of individual words. To this end, we should analyze the grammatical relationships formulated in the statement of a problem. Such work should precede the analysis of its logical and mathematical relations and be conducted at the orientation stage, which in terms of its content actually boils down to the analysis of grammatical relationships expressed in the statement.

It would be only logical to assume that difficulties with the direct understanding of logico-grammatical structures caused by a local brain lesion will result in the disordered analysis of the statement of an arithmetic problem. Therefore, patients with such injuries suffer from severely disturbed processes of understanding of the logico-grammatical structure of a problem which can easily be performed by any normal person. To illustrate this conclusion let us consider how such patients solve arithmetic problems.

The first group of patients to be carefully examined have a syndrome of marked semantic aphasia.

Patient B, 40-year-old male (case history No. 34965), a teacher, was left with severe aphasia centering on the impaired understanding of complex verbal forms (semantic aphasia) after a severe embolism in the middle cerebral artery of the left hemisphere.

His personality characteristics remained quite intact. He was critical of his condition and could correctly assess his defects. The patient was consistently working towards the recovery of the lost functions. He was in general, hard-working and resourceful.

The patient's neuropsychological status notes only the minor impairment of deep and tactile sensitivity and afferent paresis of the right arm and hand which was more pronounced in the distal sections. The patient encountered slight difficulties in *position praxis* and *spatial praxis.* He failed to directly imitate a certain spatial position of the hands. The desired result could be obtained only by extensive external supports and the verbalization of actions. The patient encountered the greatest difficulty when performing tests of *constructive activity.* While solving such problems the patient attempted to analyze them comprehensively and distinguish a series of supports, however, he made serious errors during the course of problem solving.

Speech deficits which had been quite diffuse in the early stages of the disease at the time of the assessment partly regressed and assumed the form of unmistakable semantic aphasia. The patient found it extremely difficult to deal with certain logico-grammatical constructions. For instance, he could not directly grasp the meaning implied in comparative and attributive constructs. He encountered serious difficulty in figuring out the idea of the constructions with the prepositions "from," "to," "on," "by," "under," "over," etc. To gradually comprehend their meaning he should have performed an extended series of consecutive operations every time.[1]

Reading and writing were measurably restored. But he still had some specific motor and sensory problems. His reading revealed difficulties in text comprehension caused by impaired understanding of logico-grammatical constructions.

Arithmetical operations were severely impaired. This was especially true of the operations which require carrying over from the tens column. The difficulties were due to the need for stable spatial orientation in performing such operations.

The patient's *visual gnosis* was perfectly intact. He could recognize objects, letters and geometrical figures correctly, even in sensitized tests.

The assessment showed that this patient had already begun to encounter difficulty in problem solving at the stage of *repeating the statement of a problem.* As a rule, he nearly always failed to repeat the statement of an orally presented problem since he could not immediately grasp the meaning of its grammatical constructions. It is important that despite severe speech deficits, revealed both in the comprehension and repetition of a statement, the patient always eventually understood the *final question*, although initially he was not always able to associate it with the statement of the problem. Thus, the patient encountered the greatest difficulty *decoding the statement of a problem.* If he managed to decode it, the rest was quite easy for him.

Here follows the formal protocol of a series of experiments which will give us a better insight into the structure of this patient's impairments.[2]

The experimenter reads out a simple problem: "Nick had 21 roubles and his brother had one third of the sum. They spent all their money on an aquarium. How much did it cost?"

The patient tries to repeat the problem but encounters serious difficulty "Nick... had 21 roubles . . . and his brother . . . and his brother . . . brother his . . . his . . . what's the matter: his brother? . . . there was something there . . . one third... no, something is lacking here (a pause). Aha! One third of it. They bought an aquarium. As far as I could understand all... all the money was spent... well,... how much did the aquarium cost"?!	He faced difficulty in analyzing logico-grammatical relationships repeating the statement.
And how will you solve the problem? "The problem can be solved on two operations... no, in two... on two operations. The first question is — we should... learn... the brother... how much money he has. Thus, 21 should be divided into three. It equals seven."	Searching for the correct grammatical phrasing.
And what does seven stand for? Seven . . . seven roubles . . . money . . . brother . . . the other, the younger... and the second operation... we should add the money... and the younger and the elder and it 28 roubles. *And what does it mean?* This is the . . . this is the . . . what do you call it? The change... for the aquarium, that is, the cost... that is, how much the aquarium costs."	Paraphasia. Despite the recognized logico-grammatical difficulties, the patient is working his way to the idea of the problem.

The protocol demonstrates that despite the initial difficulties caused by the defective analysis of grammatical constructions the patient eventually manages to identify the main idea of the problem.

The next illustration from the protocol shows that impaired immediate comprehension of certain logico-grammatical constructions may inhibit the solution of even the simplest arithmetic problems.

The patient is asked to solve the following problem: *A boy had 27 apples. He gave some of them away to another boy. After that he had nine apples left. How many apples did he give to the other boy?*

"I haven't grasped the interconnections. But I'll try to repeat. So, the boy had 27 apples. This is clear. And he gave to another... (a pause). There was something... some word... I am dead sure it was here... But how can I fit it in? He gave nine apples away. How many apples were there left? (a pause) No. I do remember . . . that how many apples.No, nine apples were left. How many apples did he give away? Or did he give nine apples away..then? No, I am lost."

Active attempts to solve the problem.

The patient realizes the impairments in his analysis of the logico-grammatical constructions.

Logico-grammatical difficulties led to the misunderstanding of the structure of the problem, but the patient sees it and corrects the mistake on his own.

The problem is read out once again.

"Well, yes... here it is: some of them — what does it mean? I can't figure it out. (a pause) Aha! It is simply unknown, that's all, isn't it?" *You are right.* "Then it should be this way. The boy had 27 apples. He gave..to some other boy . . . It's unknown. Left. . . he had nine apples. . .left. It is unknown how many he gave away. We should learn: how many apples he gave away. Then 27... aha... should it be minus or plus? I think, 27 – 9 = 18 apples."

The idea of the problem is now clear to him and he correctly repeats it as a result of an active search for the respective grammatical constructions.

Correct solution of the problem.

After the problem had been correctly solved but before the meaning of the construction "some of them" became absolutely clear to the patient, he was asked to think of a similar problem. He once again encountered serious difficulty in figuring out and using correctly the logico-grammatical construction which expresses definite logical relationships.

"A school bought 36 maps... and gave away to another... not to another... just a moment... May I think? (a pause) And gave away to some other... well, let it be n-maps . . . How many maps were left?(a pause) No, there is something wrong here. I'm being tortured by the middle of it. again. How many did they give away? What does it mean?!

The patient once again misunderstands logico-grammatical constructions. He attempts to word the idea of the problem in algebraic terms.

The investigator reads to him the original problem. "A school bought 36 maps and gave away to another school... the school gave away... something escapes me here. So, a school bought the maps. Then it gave away... How many? Aha! We do not know how many... What do they say in such cases?" *Some of them.* "Here it is. some of them, some of them, well, this expression isn't quite clear to me yet. Well, let it be. After this n- maps were left. We should, of course, learn how many maps were given by the first school to the second."

He sees his mistakes.

Search for the adequate language. Correct formulation of the problem after the desired logico-grammatical constructions have been found.

The formal protocol shows that the understanding of a problem's logical structure depends on primary difficulties in decoding the meaning of logico-grammatical constructions.

Such defects led to the frequent failures to grasp a problem quickly. The misunderstanding resulted in the incorrect repetition of the statement and the final question. The patient, however, always assessed his defects correctly and tried to gain insight into the problem's logical structure through deliberate detailed logical reasoning.

The patient is given the following problem to solve: *The shadow from a pencil which is 16 cm long is 48 cm longer than the pencil itself. How much longer is the shadow than the pencil.*

The patient's repeated attempts to recite the problem ended in failure: "The pencil and the shadow... longer... longer... here it is again — longer — what's that? What does it mean? I can't understand, what am I to do?" The patient gives up his attempts to repeat the problem and solve it. He asks the experimenter to give him the text of the problem for independent reading. The aphasic reads the problem very slowly, thinking over nearly every word in it. He repeats some of the constructions many times.

Logico-grammatical difficulties.

Do you remember the problem now? "Yes, I do but the point is different: how to locate all this." *Repeat the problem.* From... from... that's it (He is drawing an outward arrow): the pencil casts the shadow... yes, the shadow... by... just a moment, it seems to me I've grasped it... by, it seems to me, isn't it? 48 cm longer... no, gosh!... I have still failed to grasp something... I can't repeat it."

The extended method of decoding the grammatical structure of the problem.

Active verbal analysis the problem.

Listen once more attentively. I'll read the problem very slowly. The problem is read aloud for the third time. The patient is listening with strained attention. He tries to get the idea, moving his lips in an attempt to repeat the problem after the investigator.

Did you get the idea of the problem? "Not quite." *And do you know at least approximately how you can solve it?* "Certainly not. I don't know the problem, I didn't quite understand it. I know that there is a pencil and its shadow in it. . .They are somehow connected . . . but, unfortunately, that's all. I remember the end... to learn... the question is... how many times but I can't figure it out again. I have forgotten everything." (He is upset.)

Realization of the difficulties faced.

We see that the patient begins to solve the arithmetic problem by analyzing its grammatical structure. In doing so, he can't begin to directly solve it without decoding its logico-grammatical form which is the bearer of certain information on the relations between the given objects and the idea of the problem.

The next protocol shows that the patient had frequently attempted to solve the problem before he decoded its grammatical structure, at which time all attempts ended in failure.

The patient is given the following problem: *A shop sold 80 dresses during one day and 75 similar dresses during the next one. During the second day of sale it sold dresses for 30 roubles less than during the first one. How much money did the shop make for the two days of sale?*

The patient repeats the problem: "The shop sold... during... during... one day 80 dresses... and during... during... (thinks for a long time) during... the next. . .Well, there is something I can't figure out, it seems to me it was like this... and during the next one 75 also no, too... no, it isn't "also"... it's dresses. The second day the shop sold "for" or "in" no, it seems to be for 30 roubles less than during the first... the first... How much money did the shop make in all in, no, it should be again "for" two days... No, how much, how much there must be something wrong here."

The active search for the meaning of the prepositions.

Despite the logico-grammatical difficulties the question of the problem is repeated correctly.

Did you understand the problem? "I did and I didn't at the same time, there are so many these "for's"... there... I may have understood or misunderstood it. Let me try and see what will come out of it."

The patient begins to solve the problem. "75 + 80 = 155 dresses... these are the dresses sold by both shops... (thinks for a long time) No, it must be different, okay, I'll try this way. Then it must be as follows: 155 × 30 = 4650 roubles." *Is that all?* "Yes, that's all. But I am not sure that it's correct, you see, I didn't quite get the statement, I grasped this well as soon as I began to solve it. I should have understood something else and only then solve it, because it is, probably, wrong, isn't it?" *And what do you think?* "It's difficult for me, I know what I am to learn . . . how much money did the shops get . . . but I can't figure out what it means "during the next day", "during the second day for 30 roubles more than during the first one". We first speak about the next day and then again during the second day. I understand but it somehow doesn't get here (pointing to his head).

The realization of the difficulties and their correct assessment.

The patient is given another arithmetic problem: *A housewife uses six liters of kerosene in five days. How much will she use in 30 days?*

The aphasic faced serious difficulty in solving this problem too and had to repeatedly return to the analysis of the grammatical constructions used in its statement. The patient not infrequently tried to avoid the difficulty of resolving the grammatical "problem" or to compensate for it through an abstract formulation of its specific idea. To this end he tried to retell the problem in his own words. These attempts, however, often resulted in a refusal to solve the problem.

The patient is requested to repeat the problem. He tries to do it in the following way:

"A housewife...consumes...no, it's wrong, she does not consume, it's the other way round . . . what do we call it . . . uses this n-amount of kerosene. How much does she use in several days...no,...I repeated it incorrectly. I used different words."

The attempts to overcome the logico-grammatical difficulty retelling the problem in algebraic terms. The realization of the method's inefficiency.

Try to express the idea of the problem in short simple sentences the way you did with the story. "Okay, I'll try. The housewife . . . uses . . . kerosene." (a long pause) *How many data items were there in the problem's statement?* "There were two data items but it's very difficult for me...because they are not separately but all together, I don't know..." *Try to learn how they are interconnected in everyday life and tell it in your own words.* "Okay, I'll try. A housewife uses kerosene...She cooks one day...then n-number of days. So. (a pause) No, I cannot."

The investigator's oral instructions (assistance) do not help.

What is the question of the problem? "How much kerosene did the housewife use . . . it's difficult to say . . . in n-days, isn't it." *Did you grasp the idea of the problem, what it is all about?* "No, I didn't. I tried to but all these 'in's' . . . 'on's' . . . but I think it's like this: kerosene for N-days is known . . . and for M-days . . . (a pause). No, I'm tired." (He refuses to go on with the solution)

Despite very serious difficulties faced in analyzing the problem's verbal structure the patient understood its logical structure perfectly well.

The patient is given another problem: *Some children harvested cherries from four cherry-trees — nine kg from each — and currants from two currant-bushes — three kg from each bush. How many times more cherries than currants did they harvest?*

Repeat the problem. "I don't understand what it means from four cherries . . . no, not cherries but apple-trees. Is that right?" *It's wrong.* "Then how is it? From four, from four. (repeats the phrase several times, trying to figure it out) No, it's not clear to me." *There were four trees there . . .* "Aha, then it is clear, from four . . . there were four trees, they just grew, that's all. (laughs) I thought so too but there was something I could not understand. There was something else there . . . kilograms each . . . I could not grasp that either." *They harvested nine kg from each tree. Is this clear?* "Yes, it is. Read me the problem once again." (The problem is read for the second time. While listening to the text, the patient reconstructs its language.) "And now I'll repeat it: four trees with cherries. One tree . . . each tree . . . nine kg harvested from each tree . . . And currants — two trees, no, bushes . . . and one bush gave three kg. How many times more . . . no, there was another word there . . . *By.* That's right. By how many times more cherries than currants did they harvest? "*Solve the problem.* How many cherries did they harvest, nine plus no, it's not plus nine, 9×4 equals 36 kg of cherries." *What does 36 mean?* "Thirty-six cherries." *Does it?* "No, not cherries, these are kilometers (laughs), no, kilograms of cherries."

The alienation of the sense of the genitive case construction. The patient is perfectly aware of his difficulties.

The patient quickly develops the cue.

The simplification of the problem's logico-grammatical structure results in its correct repetition.

The general solution scheme is completely intact.

Mistakes in arithmetical operations.

Paraphasia.

"The second question: how many currants did they harvest? $3 \times 2 = 6$ kilograms of currants (with reservations) because I must multiply in this case too?" *That's right.*

"And now...how many...no, by how many this is what is not clear to me...by how many fold more cherries? 36 kilograms — six kilograms." *Did you solve it correctly?* Yes, I did (uncertainly). Or maybe I should have divided it? But I think we need subtraction here."

The defects of specific arithmetical operations: subtraction instead of division.

It is characteristic of this research that with the investigator's assistance the patient can easily solve the problem on his own, provided its grammatical structure is simplified.

The patient is given the following problem: *A quarter of a rope, which is 24 meters long, was cut off for a swing. How many meters of the rope were cut for the swing?*

The patient listened to the problem attentively, moving his lips. (a long pause)

What are you thinking about? "I doubt if I have got it correctly. It's very difficult." *Can you repeat the problem?* "Most probably, not. I haven't got the bit about the rope." *Read the problem on your own.* The patient slowly reads the problem aloud. He makes frequent pauses when he comes across certain word combinations and carefully thinks them over. On reading the problem once again he uncertainly writes down: $24 \div 4 = \ldots$, crosses it out, looks at the text of the problem and jots down the same operation again.

Above 2 minutes pass. *What are you thinking about?* "I'm doubting the operation. I didn't understand about the rope. One fourth of it?! What's that, I can't figure it out. Especially, "it" is not clear and this . . . "for" . . . the swing. Is it to make the swing? All that is a riddle to me. How it all goes together . . . It's difficult for me to understand, I don't know if I solved it correctly or not."

The same logico-grammatical difficulties.

He does not realize that he has solved the problem correctly.

We changed the language of the problem in order to make it easier for him to understand its statement. We phrased the problem in short, grammatically simple sentences accessible to the patient:

"The children bought a rope. Its length = 24 meters. They cut off a piece of it. They made the swing. The piece equals one-quarter of the length. How many meters did the children cut off?"

The patient says: "It's clear. The whole rope is 24 meters. One-quarter of it equals $24 \div 4$, the piece equals six meters."

It is noteworthy that the patient could solve a problem, even with a more complicated logical-mathematical structure, more quickly if he could work out its logico-grammatical ' constructions more easily.

The patient is given the problem: *A boy bought an ABC book and a pencil and paid 37 cents for them. Another boy bought an ABC book and two pencils and paid 49 cents. How much do an ABC book and a pen cost separately?*

The patient encountered no difficulty in solving this problem which is in logical-mathematical terms much more complicated than the previous one.

"It seems to me I've got it. It's a simple problem. All the words, all the phrases are simple. But, nevertheless, I shall try to repeat it first.

The boy bought one ABC book and one pen and will spend... spent n-amount of money... I don't remember how much exactly . . . just a moment, just a moment... aha! this is a different boy... He bought one ABC book and two pens. How much did the boy spend... no, I can't express the idea... the verb is different... spent on the ABC book and the pen separately." *How much does it cost*... That's it. How much does it cost... I have been looking for this expression, how much does the ABC book cost and how much does the pen cost. It can be easily found. I already know how to do it. We should learn about one pen... how much it costs. We should divide... not divide, simply minus... Well, how shall I put it, from the bigger the smaller and we shall learn... how much one pen costs... Then . . . we may how much cost . . . one ABC book... that is from forty... no thirty seven we must subtract... the amount... how much the pen costs."

The active search for words and logico-grammatical constructions.

The problem is repeated correctly with defective logico-grammatical constructions. The scheme of the problem is fully preserved.

The problem is solved correctly.

These protocols graphically show that, in solving an arithmetic problem, the patient first of all encounters difficulty in decoding its grammatical structure. A review of the material proves, however, that these are not the only defects patients with parietal-occipital brain lesions have since, after overcoming the difficulties decoding the problem's logico-grammatical constructions, patients encounter serious difficulty identifying the correct arithmetical operations necessary to solve the problem correctly.

For example: The patient is asked to solve the following problem: *A housewife uses six liters of kerosene in five days. How much kerosene will she use in 30 days?*

After numerous persistent but abortive attempts to get the idea of the problem the patient got special assistance after which he understood the problem, managed to repeat it, and immediately formulated the solution plan.

The repetition of the problem after the experimenter's assistance: "The housewife lives five days. Used six liters of kerosene. Lives more and

The statement is retold in simplified and extended terms.

more. We must calculate she lives, for instance, in the countryside for 30 days."

Is the problem clear to you? "Everything is clear and simple. In two operations and I could have even solved it in one operation with x. Children would do it this way: we should learn . . . this . . . how many these five, five and five there are in 30 and then we may learn how much she uses in 30 days because for one five she uses six liters. Or we shall learn first how much she uses in one day and then in 30."

The patient consciously puts forward a solution plan. Both solution methods are fully preserved.

Solve the problem. "We should first multiply 30 by five... (a pause). Just a minute, just a minute... something is wrong again. (Tries to reason consistently in detailed terms) Five... it's one time, then once more and more and all in all it will be 30 . . . Aha! we should divide . . . 30 into five, but I'm not sure. Is that right? (Looks at the investigator)" *That's right.* "The result we should, say, add. . .no, not add but multiply by six." *Solve the problem.* The patient solves the problem in writing. Then he reads it once again, compares his solution with the statement and says to himself "Correct."

The extended method of searching.

Intact control.

He faced even more obvious difficulty in finding the correct arithmetical operations while solving the next problem.

A milkmaid milked 700 liters of milk during one day, another milkmaid — 595 liters. The second milkmaid filled three milk-churns less than the first one. How many milk-churns did each milkmaid fill?

On reading the statement the patient complains of the difficulty encountered: "It's difficult for me to understand: for the second milkmaid filled . . . three milk-churns less. There were liters, why milk-churns? Maybe, it's my mistake, I may have misunderstood something."

The patient realizes the difficulties.

Having worked together with the investigator on decoding the text of the problem and the meaning of individual grammatical constructions, the patient tries to phrase the general solution plan.

He develops a vague idea of the scheme for solving the problem.

"We should first learn the difference... between this and that milkmaid, that is, the first and another, the second milkmaid. And then, probably, . . . it is a bit difficult all of a sudden . . . Maybe,

The scheme is correct but it is actually unconscious as a result of logico-grammatical learn the liters, how many liters there are in a milk-churn, what do you think? And after this we shall learn the milkmaids' milk-churns. You know, I'm saying all this but I have no definite idea of how to do it. Okay, I know for sure that the first question is 700-595 liters. The second — how many liters are there in one milk-churn? But I'm not sure How can I learn it? It's a real headache. I can't really figure out why there is a milk-churn here. I don't know if it is necessary to learn how many liters there are in it. And another problem, how to learn this. *"And if I prompt to you that there were 35 liters in a milk-churn, how will you solve the problem then?* "Then 700 + 595 = n, and then n should be multiplied by 35." *Think a bit.* "Well, there are 35 liters in one milk-churn, there are 70 liters in two of them and in three . . . what's that? I'm multiplying here, aren't I? And I must multiply then too. (a pause) No, I must divide n ÷ 35. But nevertheless I can't grasp it, it does not sink in here (pointing to his head), there is something I can't get."

difficulties. The defective search for the correct arithmetical operations.

He finds the general solution scheme. The problem is not solved as a result of defective grammatical and arithmetical operations.

The patient encountered similar difficulty in solving other problems requiring operations with numerical ratios.

The review of the above material gives us a clear idea of the difficulties faced by this patient in solving arithmetic problems. He encountered the greatest difficulty in *decoding complex logico-grammatical structures of the statement and finding the proper arithmetical operations.*

The other aspects of the patient's intellectual activity were perfectly intact:

- he remembered and understood the main question formulated in the problem;
- he attempted to orient himself in the statement and find the solution;
- the patient continuously compared the identified operations with the original statement;
- he corrected his mistakes.

Due to these specifics, the general structure of intellectual activity remained sufficiently normal despite severe defects in individual operations.

Patients with injuries to the parietal-occipital sections of the left hemisphere may suffer from disordered problem solving as a result of difficulty in decoding logico-grammatical constructions used in the statement and determining correct arithmetical operations, even if they do not suffer from marked speech deficits.

Let us analyze the following case:

Patient K, 32 years old, a teacher, studied at the university but did not graduate.

In 1955 she had an oligodendroglioma removed from the left posterior-temporal and inferior-parietal area. The tumor was located in the posterior-temporal and inferior-parietal area and grew into the inferior horn of the lateral ventricle. Her post-operation assessment noted disordered praxis of the right arm and hand (spatial and position praxis), marked oral apraxia, elements of acoustic and afferent motor aphasia. She suffered from severe impairments of spatial perception which revealed themselves in disturbed writing, reading and calculations.

In 1963 the patient was readmitted to the Institute of Neurosurgery for rehabilitation training. At that time she suffered from the residual syndrome of a higher cortical impairment caused by a parietal-occipital lesion: conspicuous defects of spatial perception accompanied by respective praxis disturbances and symptoms of semantic aphasia. Her personality was perfectly intact. She suffered greatly because of her defects, and actively sought to restore the lost functions.

In the patient's motor sphere impairments were noted that were related to disordered spatial orientation of her hands. In her constructive activity we also noted difficulties resulting from defective spatial perception.

The patient correctly fulfilled the assignments on the repetition and assessment of rhythms. We observed no optical-gnostic impairments.

Her speech did not reveal any gross defects. However, while she indicated that it was easy for her to speak in short and simple phrases, she always encountered difficulty in dealing with complex speech. She also complained of poor vocabulary "I haven't got enough words" and difficulty in formulating complex sentences. She faced similar difficulty in comprehending structurally complex speech. All her subjective complaints were confirmed by careful neuropsychological assessment.

The patient's simple, everyday speech revealed no impairments. But whenever she was asked to make an oral composition on a given topic, to retell a thematic picture or a text, to phrase a complex sentence with a given word, etc., she encountered serious difficulty.

The patient's *repetitive speech* was preserved. She could repeat sounds, syllables, words and sentences well. But with the increased volume of presentation, she exhibited mild impairments of verbal auditory memory. The patient correctly repeated a series of three to four words but failed when she was given a series composed of five elements.

The patient enjoyed intact *comprehension of directed speech.*

She suffered most of all from the defective *understanding of complex logico-grammatical constructions:* she encountered the greatest difficulty understanding constructions which in some way reflected the spatial interconnections of items or phenomena. For example, she could hardly figure out directly the meaning of such expressions as: "What comes before spring?," "What goes after winter?," "Which number comes before ten?," etc. She also had to deliberately concentrate on decoding the relationships expressed by the prepositions "under" and "over," "in" and "on," "to" and "from."

The same reason — impaired spatial synthesis — accounted for her defective calculation. The patient could perform arithmetical operations only with the help of additional extended external supports. Thus, patient K's primary impairments of higher cortical functions was spatial in nature, which resulted in a syndrome of impaired complex logico-grammatical and arithmetical operations.

This impairment had a dramatic effect on her problem solving skills. Although this patient's speech had no paraphasias, contaminations, etc., which were characteristic of the previous patient, she also suffered from defective understanding and repetition of a statement, which was mostly caused by impaired understanding of logico-grammatical relationships.

Like other patients of this group, patient K could solve arithmetic problems practically of any category (simple, complex, complex-composite, inverted, conflicting, etc.), offered to her in the experiment, which differed from each other by the increasing complexity of their psychological and logical structure. However, the critical factor was played by the grammatical structure of the problem statement.

For instance, the problem: *Victor used 27 squared copy-books and five more lined copy-books per year* is much easier in terms of its solution algorithm than the problem: *A pupil bought an ABC book and one pen and paid 37 cents for them, another pupil bought one ABC book and two pens and paid 49 cents. How much do an ABC book and a pen cost separately?* The patient, however, encountered great difficulty in determining the meaning of the grammatical construction "five more copy-books." Therefore, it took her a lot of time to solve this problem, while the first problem was quickly resolved as the patient easily understood its grammatical structure and, consequently, had no difficulty developing the solution scheme.

The patient could also solve conflicting problems. For example:

She is offered the problem: *A father is 49-years old and he is 19 years older than his son. How old are the two of them together?*

The patient tried to repeat the problem many times: "A father is 49-years old . . . the father and the son. . . may I think a bit? (a pause) It is difficult for me to harmonize everything. Let me think over the words. I've turned into a real blockhead."

Difficulties in grasping logico-grammatical structures.

Who is discussed in the problem? "The son." *What do we know about him?* "As far as I understood, he is younger, isn't he? Or not? I don't know." *And what were you to learn in the problem?* "How old together." (a pause) *Who?* "Both of them." *Solve the problem.* "You see, I do understand something. . . and at the same time not quite."

The patient understood the scheme of the problem and remembers the final question.

The protocol reveals difficulties in repeating the problem caused by an impaired understanding of the logico-grammatical structure of the statement. But the patient can, despite these difficulties, distinguish the general idea of the problem and remember the final question. Her answers, however, show that she doubts everything. Such diffidence, characteristic of all the patients of this group, is caused by an impairment of logical-grammatical relationships.

The problem was solved immediately after its statement had been reformulated into a form which was understood by the patient. For example,

I'll help you to repeat the problem. Look at the text of the problem and use it in retelling the problem. The patient gets the problem written in the following way:

The simplified method of phrasing.

A father. He is 49 years old. He has a son. His age — ? He is younger than his father. He is 19 years younger. How old are the two of them together?

The patient looks at the written statement and says aloud: The father... 49 That's clear, that's his age. It means he is 49-years old. His son is younger... aha... that's clear, his age is 19 years less. Nineteen years less. How old are they together? So,49 − 19; if he is younger then we must subtract, this is clear to me. And older, I thought I had to add. You see, I sometimes misunderstand written words. Thirty years... he, the son, is 30 years old plus 49, it makes 79. They are 79-years old together."

She is developing the strategy for solving the problem.

Difficulties in arithmetical operations.

The problem is solved.

The patient does not encounter any difficulty in resolving the problems which are even more complex in logical and mathematical terms. For instance, the problem: *The shadow from a pencil which 16 cm long is 48 cm longer than the pencil itself. How many times is the shadow longer than the pencil?* This presented enormous difficulty for the patient in terms of its logico-grammatical constructions but was easily resolved as soon as we helped her decode the difficult verbal constructs.

Here follows an extract from the respective protocol.

Repeat the problem. "From the pencil... (a long pause) It's very difficult. I don't understand anything. I can't grasp the idea... From... and then... (a long pause) Generally speaking, the pencil is so many cm. It casts a shadow... also some cm. How many times is the shadow... and what? I forgot it."

Marked difficulties in analyzing the logico-grammatical structure of the statement.

The problem is read to her again. The patient makes new attempts to repeat the statement: "The

Defective phrasing of the problem.

shadow... from the pencil... which is so many cm long... is so many times... no, so many cm more or longer than... but that is the same. So, you know I'm saying all this but I can't really understand it. How many cm, no, how many cm... times is the shadow longer?" *Do you know how to solve the problem?* "Certainly not! I haven't grasped anything yet. Let us work on the words."

Conscious attitude to the difficulty faced. The patient is aware of the reason for the difficulty.

After the logico-grammatical constructions had been simplified and the statement extended, she easily solved the problem.

"The pencil is 16 cm long. It casts the shadow. It is long... longer than the pencil... 48 cm longer. How much longer... No, how many times longer is the shadow than the pencil itself?" *Solve the problem.* "In my opinion, it's quite simple. I understood. We learn how long the shadow is. It is longer, isn't it? 16 + 48. And then the length... the shadow should be divided into . . . no, just a minute... (a pause) I'm not sure whether to divide or multiply? (a pause) Once a pencil, twice a pencil... I think I should divide the shadow into the pencil."

The patient takes up the experimenter's cue and repeats the statement correctly.

An extended method of searching for the proper operations.

The protocol shows that the patient's purposeful activity is unimpaired. She begins to solve the problem with the attempts to analyze its statement. It is not easy for her to do this, because of the defects which she actively tries to eliminate. In the process of analysis the patient develops a strategy for problem solving. She correctly determines the proper sequence of operations.

At the investigator's request the patient quickly thinks of a correct similar problem.

"Twelve kg of cucumbers were harvested from a kitchen garden... and tomatoes — 36 kg... more. (a pause) I'm checking whether the statement is correct. So. How many times less tomatoes . . . sorry, cucumbers than tomatoes were harvested? The solution is as follows . . . We should first the learn how many tomatoes were harvested and add everything to it and then... (a pause) Divide! Yes! We shall divide... tomatoes into cucumbers."

The transfer of the principle.

No difficulty faced in the course of problem solving.

The test on the independent formulation of problems after a given pattern, provided additional proof that the patient could understand logical structures. The comprehension process was intact, although it was more difficult for her to figure out their meaning because of speech deficits.

It is noteworthy that such patients generally resort to a roundabout method of phrasing the problem on their own initiative which makes it easier for them to understand its grammatical structure. They use lots of details. In their presentation the problem resembles a story.

This was very evident, for instance, when the patient was requested to think of a problem similar in type to the following one: *A son is five-years old. In 15 years the father will be three times older than the son. How old is he now?* The patient coped with the task. After several abortive attempts to formulate the problem in the standard abbreviated form, she resorted to the extended method of structuring the statement. Therefore, it was easier for her to express the logical relationships implied in the problem in the proper grammatical terms.

"A grandmother is 60-years old . . . And her granddaughter... her granddaughter is five times older. How old is the granddaughter? (a pause) Oh, no! What's the matter with me?! The problem is absolutely different and it's all wrong altogether. I wanted to make it shorter, the way it should be. But it's difficult for me to get who is older and who is younger It just seemed to me for a moment that I grasped something-and everything is gone. I'll tell it with more details, it is clearer for me this way. The granddaughter is five-years old. Her granny is older. Years pass. 20 years have passed. The granny is now four times older than her granddaughter. How old is the granny now?" *How will you solve the problem?* "Well, it is quite clear. Five years plus 20. It gives 25-years. This is the granddaughter's age. And her granny is four times older. We must divide..., no $25 \times 4 = 100$ years. And now — $100 - 20 = 80$ years."

Abortive attempts to formulate the problem in the standard grammatical form.

Conscious attitude to her mistakes, search for the roundabout ways.

The simplified (extended) form of formulating the problem.

Perfectly intact solution scheme.

When this patient began to resolve the problem directly, she faced difficulty in searching for the correct arithmetic operations caused by impaired simultaneous synthesis, which was similar to that which was encountered by the previous patient. We are not going to discuss these impairments in detail now. Our objective is to show, using this case as an illustration, the nature of the impairment of arithmetic skills in this group of patients for whom it was once a highly automatized ability. In parietal-occipital brain lesions, the disintegration of this skill turns it into a series successive individual operations which require external support. This method of performing arithmetical operations essentially reveals the internal structure of an arithmetic operation in the non-contracted and unmastered form. The simultaneous understanding of the statement disintegrates into a series of successive individual operations, and therefore it becomes difficult for the patient to solve the problem.

The patient is given the problem: *A housewife uses six liters of kerosene in five days. How many liters will she use in 30 days?*

After the patient had solved the problem in one possible way, she was requested to solve it differently.

"I'll think." *Read the statement attentively, analyze it again.* "So. She uses in five days. . .and then in five days more . . . five plus five and six plus six, so she'll use twelve liters in ten days. Now once again plus five but this time we should add already to ten. . .and plus six. . .no, I'm all muddled up." *How many times are you going to add?* Until it makes 30 days. *And how can you learn the same without adding the fives?* "I don't know (a pause). Maybe, by multiplying them, is that so?" *No.* "Well, then by dividing 30 into five. . .I don't know how much that will be." *That will be six.* "Then, $6 \times 6 = 36$. A housewife will use 36 liters of kerosene in one month. But you know, I'm doing it but it is very difficult for me to understand all this."

General strategy is intact. A good idea of the inner meaning of the multiplication operation. She moves to multiplication through addition. She does not fully understand the meaning of arithmetical operations and their interconnection with logical relationships.

The patient performed the operation of division in a similar extended way. While performing the operation, she used external supports which helped her to understand and master the internal meaning of division. For example, she divided 64 into 16 in the following way:

"Sixty-four should be divided into 16. What does this mean? Aha! 16-once, 16 - twice, 16 - three times (putting out the sticks) and all this is in 64. . . How many times - one, two, three, That's it! Now it is clear."

All the patients of the group discussed actually reveal a similar disorder of arithmetical operations, although usually these impairments are less conspicuous.

This case reveals that the relative absence of speech deficits in patients with impaired simultaneous synthesis provides little insight of the essential structure of their impaired arithmetical operations.

In all such cases, the central defect of problem solving lies in the operational element of intellectual activity and is caused by the impaired logico-grammatical and arithmetical operations.

Defects in the analysis of a problem statement are even more prominent when patients with left parietal-occipital lesions also have *acoustico-mnemic impairments* which prevent them from remembering the components of the statement. It is important, however, that in these cases too, the overall structure of purposeful intellectual activity remains intact. Such patients suffer from disorders that do not go beyond the operational aspect of intellectual processes.

We shall briefly review the process of problem solving in this group of patients, using a typical case as an illustration.

Patient K, 35-years old, (case history No. 36,409) a metal worker with five years of schooling was admitted to the Institute of Neurosurgery in February, 1963 with residual symptoms of a gunshot wound which had penetrated the left parietal-occipital brain area in 1944. The gunshot resulted in total sensory and amnestic aphasia and the syndrome of spatial disorder. The latter was revealed during tasks of writing, calculation, gnosis and praxis. He gradually recovered the lost skills and at the time of our assessment we noted the following clinical picture:

The patient did not exhibit any significant defects in the elementary visual gnosis and praxis. But he encountered serious difficulty whenever he performed tests on spatial gnosis and praxis. He had difficulty analyzing the spatial relations of any elements presented to him. For example, he confused the position of the clock hands and could not tell the time. He failed to read a geographical map or perform the tasks of constructive praxis.

His verbal activity was sufficiently preserved but he had difficulty in naming objects, repeating long word series (he complains that he can't remember them, "the words go away") and understanding complex logico-grammatical constructions. He also had serious difficulty performing arithmetic operations as a result of defective analysis of the place digit structure of a number and disordered operations with numerical ratios. He also was noted to forget words, which is a sign of the syndrome of acoustico-mnestic aphasia.

The assessment of his problem solving process revealed a clinical picture similar to the previous patient described above. But he also exhibited marked acoustico-mnestic deficits which manifested themselves by his inability to memorize a statement. Each of this patient's attempts to eliminate difficulties in remembering and reproducing a problem on his own resulted in failure. As a result, he often had to give up his attempts to repeat the problem aloud.

With a view to eliminating these impairments, the patient was given the *written* text of a problem for independent reading and to facilitate its memorization. The patient usually read it aloud many times, then repeated it several times: first in parts and then the whole problem. But our investigation showed that such work did not eliminate his problem solving difficulties. Thus, by reading a problem and trying to understand and repeat its statement, this patient faced all the difficulties encountered by the patients described above. Acoustico-mnestic defects aggravated the general picture of impaired problem solving but did not alter its character. For example:

The patient was given the problem: *A boy had 24 apples. He gave some of them to another boy and had seven apples left. How many apples did he give away?*

The attempts to repeat the audited statement end in failure: A boy had... 24?... yes, it seems to me 24 apples... And then... no, I can't get anything. I did not catch on. Read it again, please." The problem is read for the second time.

Difficulties in retaining the statement of the problem.

The patient immediately repeats it in parts after the investigator but once again fails to repeat the whole problem.

"The boy had 24 apples . . . He gave to another boy . . . No, may I read it?"

The patient is given the text of the problem which he slowly reads many times, pronouncing aloud some constructions in the statement which he could not figure out. "Okay, I'll try. A boy had 24 apples. He gave some of them to another. How many left? . . . No, I know seven . . . but I don't remember what happened to these seven. . .I'll read once more. So,a boy had 24 apples. He gave some . . . some of them to another boy. How many left . . . How many apples . . . seven left . . . No, there is something wrong . . . there was seven there . . . How many apples were left? No, how many apples did he give to another boy, but there was a seven there. The moment I look away from the text I can't figure out anything."

Logico-grammatical difficulties. The patient did not grasp the meaning of the construction "some of them" and encounters difficulty in memorizing the statement.

Read the text of the problem and solve it. The patient reads the text and says: "But I'm reading and can't get what it is all about. What does "some of them" mean? I've got some vague idea but I can't visualize it."

After we had worked on the understanding of the text of the problem the patient cried out: "Gosh! But it's all so clear and so easy, $24 - 7 = 17$ apples the boy gave away. Certainly, there were 24 apples. It is unknown how many he gave away. Seven were left. How many did he give away — that is the question."

Due to the elimination of difficulties in the verbal analysis the patient could solve the problem correctly and consciously repeat the statement.

The patient could quickly and easily solve similar problems if their grammatical structure was clear and understood by him. For example, he solved the following problem correctly, as soon as he had read it. *Some birds were sitting on a tree. Then 28 more birds came and all in all there were 54 birds. How many birds were there initially on the tree?* "Fifty four minus 28 . . . that will be so . . . 34 minus four and four. . . Twenty six birds." He did not encounter any serious difficulty in repeating the statement either.

Correct solution.

Repeat the problem, please. "May I read it? So. (after he read it twice) Some birds were sitting... Unknown... 28 more came... and sat on the tree and all in all there were 54 birds. And how many birds were sitting at first — is unknown. So, how many birds were there at first?"

These illustrations show that the difficulty in solving the first problem was also caused by the defective understanding of logico-grammatical constructions.

Our experience proves that when these difficulties were eliminated, the patient could solve all types of problems: conflicting, standard, indirect, inverted, etc. But all of them could only be solved due to an extended orientation activity. Thus, the patient could develop a general solution strategy only having performed a detailed series of orientation operations.

The patient is given the text of the problem: *A father is 45. He is 19-years older than his son. How old are the two of them together?*

Manifest logico-grammatical difficulties in repetition: search for proper prepositions and constructions.

The patient reads the problem attentively and tries to repeat its statement: "A father is 45... the son... the boy... of the father by 19... no, 19. How old are they together? I repeated but it seems to me I didn't grasp it."

He is not sure whether he has repeated it correctly.

Tell me what is your solution plan. "A father is 45... Who else is there? Aha!... the son... I forgot... (reads the problem again) So, we should first subtract... and then... then..."

The patient attempts to overcome the acoustico-amnestic difficulties; after that he managed to develop the solution strategy.

What will you learn if you subtract? In my opinion, how old the son is... It's difficult for me to figure out what it means... (reads aloud) "older than his son". So, the son is young, the father is old. Well, that's it, to subtract and then add the years to the result."

In this test, the patient reveals not only acoustico-mnemic difficulties but also manifests logico-grammatical defects, which he actively tries to overcome. Once they are eliminated, it becomes clear that the general solution plan is completely preserved.

The additional difficulties faced by the patient were caused by acoustico-mnestic impairments which were especially evident on tasks involving the *independent formulation of a problem on a given model.* The patient altogether failed to orally formulate a similar problem without the model's support (retaining the given model in memory). Whenever the subject was allowed to write down the formulated problem using the text of the model-problem, the difficulties which were caused by his acoustic-mnestic deficits disappeared, but the difficulties typical of all patients with parietal-occipital injuries to the left hemisphere persisted. The experiment showed that the patient always omitted the part of the problem worded in a way that was misunderstood and, therefore, inaccessible to him. Like other patients of this group, he often resorted to a detailed narrative style for formulating the problem statement, using simple grammatical constructions.

For example, the patient was given a model-problem (which was taken away after he had read it many times): *There were 38 small and 27 large berries in the basket. Some of them were taken by the children. After this, 45 berries were left. How many berries did the children take?* Using this problem as a pattern he thought of the following problem: "I went shopping and bought red and black prunes. I counted how many prunes I had bought. There were 43 black prunes and 38 red ones. A friend of mine came up to me and I gave him . . . treated him . . . gave him 18 prunes. I should learn how many prunes were left. "When the investigator asked the patient if he had performed the task But when he directly com-correctly, the patient hesitated, it seemed to him that "there was something wrong" but he couldn't say what it was. When the patient compared the formulated problem with the model he immediately found the mistake and corrected it. What's the matter with me? This is precisely what I doubted . . . "some of them" . . . I had to say . . . I gave him the unknown number . . . And then . . . Eighteen prunes were left. I should have remembered how many prunes I gave him."

An extended form of phrasing.

The problem's logical structure is changed. The most essential part of the problem formulated in the model as "some of them" was omitted.

But when he directly compared his statement with the model, the defects, typical of the work from memory, disappeared.

Therefore, we see that this particular patient differed from other patients of this group by impairments of verbal memory which were part of the syndrome of acoustico-mnestic aphasia; in solving arithmetic problems these defects were easily eliminated through providing certain conditions.

The material accumulated in the course of the experiment gives us every reason to assume that the grammatical structure of a problem statement is very important to its eventual solution. This conclusion was graphically illustrated by all the patients in the study. Each patient with parietal-occipital lesions, without exception, encountered serious difficulties that often could not be overcome when analyzing a problem statement, despite the fact that their active searching activity was perfectly preserved. Their difficulties understanding the meaning of logico-grammatical relationships were, of course, caused by defects of the verbal system which, in turn, were the result of impaired simultaneous spatial synthesis. This disorder accounted for the severe disturbances in arithmetical operations seen in our patients.

Thus, in patients with parietal-occipital lesions, the impaired analysis of logico-grammatical relationships and impaired performance of respective arithmetical operations, described above, were the chief causes of impaired intellectual activity.

These defects were particularly observed against a background of intact attention, orientation toward the ultimate goal, and selective behavior characteristic of all the patients of this group. We should note that these patients also preserved a critical attitude to the results of their activity and their impairments. When attempting to solve a problem, such patients continuously compared their operations with its statement, noticed and corrected their mistakes and responded quite emotionally to the difficulty encountered when they attempted to identify the meaning of logico-grammatical relationships and determine adequate problem solving approaches. Such intactness of the basic components of intellectual activity-orientation in the statement, development of the solution plan and continuous control over the solution process — provided for the correct solution whenever the problem's logico-grammatical structure could be simplified.

This experience shows that all patients in this group encountered similar difficulty in grasping the logico-grammatical structure of problem solving.[3] We did not come across a single case of impairment in the organized purposeful intellectual activity and uncritical attitude to one's intellectual operations which is typical of patients with injuries to the temporal lobes.

Problem Solving Training of Patients with Parietal-Occipital Lesions

The specific disorders of the problem solving process in parietal-occipital lesions, described above, were especially evident in pilot *rehabilitation training* of such patients.

In our studies we proceeded from the assumption that the use of a system of auxiliary techniques, designed to help our patients grasp the logico-grammatical structure of a problem, would provide a certain compensation for these defects and partial restoration of their impaired intellectual operations.

In the first stages of training, our patients were taught to use some techniques that enabled them to distinguish the significant elements of the problem statement and substitute a detailed process of analyzing the logical elements of the statement for the direct comprehension of its logico-grammatical relationships. Patients were asked to underline, while reading the text of the problem, any sentences which were unclear to them and instructed to rephrase them, using simpler language. Then they were asked to restate the problem in their own words and develop a plan for its solution.

After they had carried out these instructions they were given "a glossary" of common grammatical constructions and the corresponding arithmetical operations:

"By how many units more"	we should subtract (−)
"By how many units less"	also subtract (−)
"n-times more"	multiply (×)
"n-times less"	divide (÷), etc.

This method yielded certain positive results, but unfortunately any progress made was transitory in character. For example, patients could not follow long, spoken instructions (or a program) and often turned to the investigator for help. Focusing attention to the first part of the instruction, the patients forgot the rest of the operation plan, causing the investigator to repeat the instructions many times. We should note that patients with acoustico-mnemic disorders could not use this system of training at all. To eliminate these difficulties, we divided the program into a series of separate instructions which were consecutively presented to the patients after they had completed the previous instruction. This made it somewhat easier for them to understand the statement of a problem.

We made significant headway when we began to present all the instructions (including the methods for performing the required operations) in strict sequence. To this end, we developed a special training program. Its content was reevaluated after every experiment but its structure remained unchanged. The program consisted of a number of instructions to successively perform specific operations that promoted the analysis of problem statements and the identification of effective arithmetical operations. All the instructions and a detailed list of consecutive operations necessary to solve the problem were presented to each patient. The program looked as follows:

1. Find and underline the main words of the first sentence.
2. Underline *what* is said about the *main object* in this sentence.
3. Write down *schematically* the idea of the first sentence.
4. Retell the first sentence in your own words.
5. Find and underline the *main words of the second sentence.*
6. Underline *what* is said about the *main object* in the second sentence.
7. Write down *schematically* the idea of the second sentence.
8. Retell in your own words what you have understood.
9. Retell the text of the problem. Use short and simple sentences.
10. Solve the problem.
11. Please, remember that:
    - "How many times more (or less). . ." = ÷
    - "How much more (or less)" = ×
    - "N-times more (older, longer, etc.)" = ×
    - "N-times less (younger, shorter,etc.)" = ÷
    - "N-units more (kg, m, l, etc)" = +
    - "N-units less (kg, m, l, etc.)" = −
    - "By so many units" = disregard the preposition "by."

This program externalized the underlying problem solving operations and attempted to compensate for *impairments which are most disordered in patients of the group under study.* It provided an extended and successive process of translating logico-grammatical constructs into arithmetical operations. The program was not designed to organize the patient's behavior while he analyzes the logical structure of a problem: we have already noted that the behavior of such patients remains intact and its restoration does not require any additional efforts. The program focuses instead, on a required sequence of operations related to the analysis of a problem's grammatical structure. This program enabled the patients to perform problem solving tasks that they previously failed as a result of their respective impairments. It was quite successful because it separated an integrated process of perceiving and understanding into a series of successive deliberate operations. Furthermore, these operations were first externalized and materialized, and only later were they gradually eliminated.

For example, the following illustrations show how the patients mastered the process of problem solving with the help of the suggested program of operations.

Patient B (the information on the case was provided above) was given the problem: *There were 18 books on two bookshelves. There were two more books on one of them than on the other. How many books were there on each shelf?*

The patient asked the experimenter to repeat the problem. "On each shelf... no,... not on each... how do we say that there were 18 books altogether? The first... of... them... What does it mean "of them"? It's difficult for me... I can't figure out... how many shelves and... how are they?"

The patient is given the program. He works slowly on each sentence resorting to the investigator's assistance. "So. There are 18 books on two bookshelves. The books and the shelves are the essentials. What is said about them — they have 18 books. Let's go further. On one of them — there are two more books. This is the riddle. So." (thinks for a very long time) *Insert the omitted word. Which word is missing here?* (a pause) "The shelf, aha. On one shelf... shelf... this is the essential point. Then what is said about it... aha!... there are two more books... more — is that correct? (draws the sign >).

Now the whole of it. On one... shelf there are more... as compared... with this (points to the first sentence) by... for... no." *By how many?* "By two books." Writes down:

1 sh. + 2 sh. = 18 books
1 sh. > than 2 sh. by 2 books + 2 books.

"How many books have the first and the second shelf separately, plus two books as there are two books more, isn't it? That's it." (Consults the program) *Retell what you have grasped in your own words.* "I've already retold it. Here (draws) = = = = = = = = = = = = = 2 shelves. Let's say there are some books in a book-case-n-books. One shelf — this one is bigger... in... it, by two, well, let it be by n, how many books are here and there?"

(Items Nine and Ten of the program turned out to be redundant, they were not informative. The patients usually moved to retelling a problem and therefore these items were later struck out of the program.) "So, it's like this. (consults the program) There are two shelves in the bookcase. On... in... them there are books... 18 pieces. One shelf and the other shelf differently. One... on one shelf there are more... two more books. How many books did this and that shelf have? (reads out: *now solve the problem.*) We may do it this way. (a pause) There are altogether 18 books here... more and here everything is unknown. Aha, we may do like children do, divide the books into two. And then... and then... divide the result... no, not divide but, how is it?, minus one and we shall learn... (a pause). No, there is no logic in it as if I were trying to adjust it to the answer. We'd better do it this way. We'll subtract the extra books, minus, and the rest we shall halve. And then we'll add the extra books again... not add but increase:

$18 - 2 = 16; 16 \div 2 = 8; 8 + 2 = 10$. That's all."

The patient is given the problem: *A son worked 23 workday units (units of work on Soviet collective farms), his mother - nine workday units more than her son, his father—19 workday units less than his son and wife taken together. How many workday units did the father work?*

He tries to repeat it: "The son worked... 23 work... what do they call it... in collective farms... well, work... units... This problem is in general difficult for me... "more", "less"... and besides, there is here... I always fail to figure it out straight away. I must think. (a pause) The son earned... no, worked[4]... how shall I put it?... he worked so many , don't remember exactly, workday units, and his mother? (a pause) As for his mother, it's not simple... it's complicated for some reason and it's even more complicated for the father—both the son and the mother. No, I misunderstood something. Read it once more, please. And better give me the table (the program) and let us work and not repeat." (laughs)

The patient is given the program. He consults it and successively fulfills each instruction. "So. The first sentence is simple. A son worked 23 workday units. The son worked — this is the main point, and 23 — this is about the main point. His mother — nine workday units more than the son. His mother is the main element and (a pause) son... son... son... no." *Insert the omitted word.* "Mother... but I don't know what is omitted here." *Which are the main words in the first sentence?* "The son earned." *And what about his mother?* "Aha, she also worked—this is the essential point, more (draws the sign >), is that correct? So, more, nine workday units more. And what about the son[5]... What's that?" *Think. She worked more. This is a comparison with something, isn't it.* "I realize that this is the comparison, but the son... I can't understand this word." *Then let us put it this way: The mother worked nine workday units more compared with...* "Yes, that's it, certainly, than the son," etc.

The patient drew the following scheme of the problem:

Son (s) = 23 units

Mother (m) = > 9 units... + 9 units than her son

Father (f) = 19 units... than M + S

How many workday units did the father work?

While solving the problem the patient used the scheme, although he resorted to the detailed written program all the time in order to check himself — "more — shall we add or subtract," etc. The problem was solved correctly.

The illustrations show that this program is certainly of great help for training such patients. It can be mastered easily. The patient used the program: he first read each item and then it was enough for him to recite it. In the most difficult cases he returned to the written program in order to make sure that he performed the operation correctly or find the mistake. But the written program was used exclusively for control purposes.

We had similar results while working with other patients in this group.

Patient K. (information on the case was given above) received the problem: *Some workers lay five pipes in four days. How many pipes will they lay in 16 days?*

At first the patient could not repeat the problem. She complained of poor memory. She managed to repeat the whole text only after the problem had been read to her in parts.

After this the patient attempted to formulate the solution plan: "I somehow fail to grasp the connection. . . well, should I learn in one day and then in 16. . . or. . . I don't know anything else or how to do it, my head does not work."

Try to recollect how we've solved such problems. "I remember — with fractions but it is hard for me to grasp how I can find them. I might be able to recollect if I try really hard but I want to do it on my own and I can't." *Try to solve it on your own then.* (a pause) "Let us take four days for one fraction (a pause) and what then? And then I don't understand anything. In my head everything is separately, nothing matches." (she begins to cry)

The patient is given the program. She reads it attentively and attempts to analyze the statement. "In the first sentence the main word is — some workers. They lay pipes . . . five pipes . . . in four . . . in four days . . . well, that's clear. One day, two . . . days, if we add everything it'll amount to five pipes. Schematically, it will look as follows: 4 days = 5 pipes . . . Then . . . 16 days passed . . . 4 days + 5 pipes and 16 days = x? It's all clear to me but how can I do it . . . (reads the second part of the program) How many times more. . . less. . . I think that'll do here. . . How many times more are 16 days . . . we should do as it is written here (consults the program), we should divide 16 ÷ 4 = 4. . . Four times more pipes they will lay. Then how many pipes . . . for one four of days—five pipes . . . and four more plus four more. . . (consults the table and reads: "n-times more. . . "). Probably, it should be 4 × 5 = 20 pipes . . . as it is four times more we must then multiply. . . But nevertheless, it does not sink in — all these: to multiply, to divide. . . "

The patient received the problem: *Last year Nick used 27 squared copy-books and five more lined copy-books than squared copy-books. How many copy-books did he use altogether?* He failed to solve it in one go. The support of the program yielded the following results.

"I can't repeat the problem the way it is. Let's do it in parts, in succession.(consults the program) So, the essential point here is Nick. He wrote. He used... 27 squared copy-books. Let's go on. He also used copy-books for writing... and here... it is unknown... or five more copy-books... so, (writes) if there were 27 squared copy-books, then... lined... 27... (consults the program)... aha, so, it will be 27 + 5. And then... we shall add all... 27 + 27 + 5..."

The quoted records show that the patient encountered serious difficulty, both while analyzing the statement of the problem and looking for the proper arithmetic operations. The program helped the patient to overcome her impairments unaided. Later, the patient mastered the method of work prescribed by the program and turned to it only when she searched for the desired arithmetical operations. We did not have to remind patient K., like other patients in the group under review, of the method of work to be used in the course of problem solving.

Hence, the experiments with programmed training confirmed our assumption that in the case of parietal-occipital lesions, intellectual activity is impaired only in its operational element. The impairment reveals itself in difficulty faced by such patients in grasping logico-grammatical relationships and taking a text in, as an integrated coherent entity as well as in the difficulty with determining the proper arithmetical operations. All these disorders are either directly or indirectly related to the specific disturbances of the verbal system, which take root in impaired simultaneous (spatial) synthesis.

The program offered to the patients eliminated these difficulties to a great extent and enabled them to solve the problems. Although this program did not completely answer the question of restoring the comprehension of logico-grammatical structures, it contributed to the elimination of the basic impairments in the verbal analysis and synthesis of the statement of arithmetic problems.

---

[1]Difficulties faced by this patient are essentially similar to those earlier described by one of the authors (see: Luria, 1945, 1947, 1962).

[2]The experimenter's words in the protocol are in italics.

[3]See the tables in the appendix to the book.

[4]In Russian the verbs to earn and to work are single-root words: "*zarabotal — rabotal*." Translator's note.

[5]In Russian this word has the ending "sin — sin*a*" of the genitive case which is the stumbling block for the patient. Translator's note.

Chapter 3

DISORDERS OF PROBLEM SOLVING IN CASES OF THE BASAL-FRONTAL SYNDROME

We've described the impairment of intellectual processes in the lesion of "gnostic" (parietal-occipital) brain areas. Now let us analyze the changes in the structure of intellectual activity in lesions of the frontal lobes.

Our research findings (see: Luria, 1962, 1963) show that the role of the frontal lobes in the organization of higher cortical functions differs considerably from that of the parietal-occipital sections discussed in the previous chapter.

The function of the parietal-occipital sections of the cortex, which are the central "terminal" for feedback information, is to analyze visual, tactile and vestibular information from the environment and the position of the body as well as to process it. The functions of the frontal lobes are quite different.

According to the latest research findings, the frontal lobes, which have been long believed to be the "mute zones" of the cortex, are of utmost importance for organizing mental activity and provide for the most complex forms of behavior regulation.

The frontal lobes are closely related to the reticular formation of the brain stem (truncus cerebri) and actively participate in retaining the cortical tone, which is essential for the normal actualization of the highest forms of cortical activity. They superimpose in the phylogeny over *the motor sections of the cortex* and provide the most complicated forms of integrated motor action, as they control it through the definite set of programs of operations, which are generally referred to as preliminary, "pre-start" afferentations (Anokhin, 1949). In humans these afferentations take the form of *intentions,* which determine the orientation of purposeful activity. In close interaction with the anterior areas of the *speech zone,* the frontal lobes provide the possibility of regulating man's mental activity *through the involvement of inner speech.* Therefore, they provide for the highest organizational forms of intellectual activity (Luria, 1963). Hence, we may assume that the frontal lobes are important, not so much for the reception and processing of external information, as for the *development and retention of certain programs of operation,* because they provide for the plastic change of these programs with the change in the situation and the mobile and selective character of the activity (K. Pribram, 1959).

Therefore, it is clear that an injury to the frontal lobes does not cause either visual or hearing disorders. It neither results in the impairment of simultaneous (spatial) synthesis, nor affects the ability to perform the complex logico-grammatical operations whose dysfunctions have been discussed in the previous chapter.

Our research findings reviewed in another publication[1] show that the impairments caused by a lesion of the frontal lobes are manifested even in the actualization of relatively simple *motor actions.* Whereas it is quite easy for such patients to perform simple movements either on the direct order ("raise your hand" or "bang on the table") or imitating the investigator's movements, they encounter serious difficulty in organizing the motor action whenever such a movement requires the preliminary *recoding of the signal* or is directed by a definite relative scheme.

For instance, if a patient with a lesion of the frontal lobes must raise his right hand when he hears one tap and his left hand when he hears two taps (so-called "selective response") he easily begins to disobey this conventional instruction and slides to more direct or primitive actions, or impulsively imitates the signal (raises his hand twice in response to two strokes). He may also raise alternatively his right and left hand, irrespective of the signal, reproducing the developed stereotype.

The impaired ability to act within the framework of a given scheme and the tendency to slide into more primitive forms of action (the echopraxia of a model or the inert reproduction of stereotypes) are the basic effects of an injury to the frontal lobes, which deprives the behavior of such patients of organized selectivity. Therefore, the disturbance of the "psycho-regulatory function," first mentioned by Bekhterev, is one of the central symptoms of an injury to the frontal lobes.

It goes without saying, that the inability to control one's activity by certain programs as a result of frontal lesions is not limited to relatively simple movements, but drastically affects the realization of *intellectual activity.* Therefore, we have every reason to assume that patients with frontal lesions suffer from severe impairment of the problem solving process. This impairment will, however, dramatically differ in character from the disorder of problem solving in cases of parietal-occipital lesions.

Prior to discussing our specific research findings we would like to touch upon one important issue. It is common knowledge that lesions of the frontal lobes may take place against different backgrounds: on a background of *irritation,* on the one hand, and on that of *inhibition* accompanied by inactivity and pathological inertness, on the other. Therefore, the clinical picture of the cases under review may be completely different.

As noted above, the frontal lobes of the brain are not always a homogeneous entity. Clinical investigations have long pointed to the different nature of disorders caused by an injury to the basal posterior-frontal and prefrontal sections of the lobes, although in all the cases, the resulting impairments cause the aforementioned structural impairments of purposeful intellectual activity.

Hence, it is preferable to discuss the forms of problem solving disturbances caused by different types of the frontal syndrome using as an illustration one patient typical of the given group.

The first group of patients with disordered intellectual dynamics consists of people with injured frontal lobes who, in principle, are able to properly realize intellectual activity but fail to do so as a result of pathological irritability which easily disturbs the normal actualization of intellectual processes. Patients with an injury to the *basal areas of the frontal lobes* are typical of this group. It is noteworthy that the dysfunction of the basal section of the frontal lobes may be caused both by a pathological process at the base of the brain and a pathological process (for instance, a tumor) located intracerebrally—in the white matter of the basal areas of the frontal lobe — which affects, in the first place, the normal functioning of the basal sections of the frontal lobes.

The intellectual impairments observed in these cases usually occur against a background of behavioral changes that are well known to clinicians. They usually reveal themselves in the general disinhibition and impulsiveness of such patients who easily lose control over their behavior and in frequent affective outbursts which they can barely rectify. The intellectual processes of such patients may remain relatively intact: they solve simple problems easily as they are able to grasp the main idea and develop adequate solution schemes. In the case of a slightly more difficult problem, however, they often omit the necessary stage of orientation in the statement and simplify the statement in the course of retelling it. They sometimes substitute a fragmentary answer to the partial question for the complex hierarchy of operations, or replace the system of organized logical reasoning with the stereotyped repetition of previous answers.

This group of patients are, however, characterized by the fact that the impairment can be easily compensated for, and normal intellectual activity restored by inhibiting impulsive answers (when the patient is requested to repeat in detail the statement of a problem) and consolidating the orientation basis for intellectual activity.

Therefore, the defects of problem solving observed in this group of patients often resemble the mistakes made by normal subjects who solve a given problem when they are pressed for time or extremely tired, and thereby lose the necessary control over their behavior.

We shall illustrate our analysis of intellectual disorder, typical of this group of patients, by the review of a case where the symptoms of intellectual dysfunction gradually became increasingly more evident, as a result of the growth of an intracerebral tumor in the basal section of the frontal lobe. Therefore, in this case we could observe the dynamics of intellectual dysfunction from its minor manifestations to the most severe forms.

Patient B: General information

The patient is 42 years old (medical history No. 35,070). She was admitted to the Institute of Neurosurgery, complaining of headaches accompanied by vomiting, general weakness and infrequent attacks attended by loss of consciousness and vegetative symptoms but without any aura.

The objective assessment noted the following picture: the patient is inhibited, she does not perform the tasks at once, she is inactive, unconcerned about her condition and does not show any emotional responses either to her condition or the forthcoming operation. She cannot coherently retell her case history and shows marked symptoms of mental disorder related to mnemic deficits, evident affective changes and fluctuations in the condition of her consciousness. She is insufficiently critical of her condition. The patient believes that her memory is as good as it was and she could have successfully performed her chores at work.

Her neurological status notes marked papilledema, bilateral pathologic reflexes which are greater on the right, and paresis of upward gaze. Her sense of smell was intact. We observed no sensory deficits.

Her EEG revealed focal abnormalities in the left frontal-temporal area against a background of severe diffuse brain dysfunction. Her arteriogram shows the dextroposition of the left anterior cerebral artery, which is indicative of an intracerebral tumor in the left frontal lobe.

Neuropsychological assessment revealed that all forms of praxis, including dynamic praxis and reciprocal coordination, were fully preserved. The patient could correctly evaluate and reproduce rhythms, both shown and verbally described to her. We observed no symptoms of the pathological inertness of established stereotypes. She did not exhibit any disturbance in the formation of selective responses to a verbal instruction. The developed stereotyped sequence of reactions is easily broken, in accordance with the changing signal sequence. The only pathological symptom in this case, is a minor tendency toward premature impulsive motor responses.

We did not observe any gnostic defects. Her speech was intact. When repeating a series of five words, the patient changed the word sequence and showed some symptoms of persevering with the wrong word order which she had established. She could easily perform simple arithmetical operations — counted backward from a hundred in sevens making very few mistakes. She encountered no difficulty in assessing logical-grammatical relationships. The patient lost the necessary selective connections and regressed to by-associations only in the tests of analogies, but often corrected her mistakes unaided. She could determine the idea of a narrative picture or a story adequately enough. Although she sometimes interpreted some fragments incorrectly, making errors in her reasoning, she could, however, easily correct these mistakes if she were asked some additional questions.

The patient suffered from a psychological disorder with marked symptoms of loss of spontaneity and inactivity, emotional changes, and an uncritical attitude to her condition, against a background of relatively intact intellectual operations. These mental dysfunctions suggested a tumor in the basal section of the left frontal lobe. The attacks, accompanied by the fluctuations of the tone and catalepsy as well as the preserved sense of smell supported the idea of a deeply located tumor, adjacent to the walls of the ventricles. This suggestion was also supported by the absence of speech disorders against a background of the fluctuating cortical tension. This preliminary diagnosis was also confirmed by the results of the EEG and the contrast methods of assessment.

The patient was operated on by Dr. Alekseyeva on January 25, 1963. During the operation the surgeon found a tumor, which grew through the entire cortex of the basal sections of the left frontal lobe. The tumor must have penetrated the anterior horn of the lateral ventricle and extended to the pole of the left frontal lobe. Its posterior penetrated the anterior regions of the left temporal convolution. The deep inner part of the tumor and its basal section were bright red with symptoms of degeneration. Histologically, this was an astrocytoma.

In this case we should distinguish two stages in the investigation of problem solving: *the first period* — before the operation and immediately after it, when the patient suffered only from minor intellectual disorders and *the second period* when all the described symptoms became much more severe, as a result of the tumor's continual growth.

Comparative analysis of the data received in both stages gave us a perfect chance to observe the dynamics of increasing disorder in intellectual operations.

Let us discuss consecutively the data accumulated during both periods of the experiment.

The Disorder of Problem Solving (The Case of Patient B)

The first stage of the disease.

The patient's condition in the first period was relatively good, so we could systematically study the way she solved increasingly more difficult problems and, thereby, describe specific intellectual processes typical of the group of patients presently under review.

It was quite easy to persuade the patient to solve given problems. If she did not encounter any serious difficulty in solving a problem she was quite willing to work on the next one. But if she faced some difficulty in problem solving, she either began to solve the problem impulsively or refused to go on working altogether, saying that she was too old for solving arithmetic problems for schoolchildren. The analysis of her problem solving process allowed us to describe her intellectual deficits, which we are going to review hereafter in detail.

(1) The patient easily repeated and solved *simple and elementary composite problems* (of the following type: $a + b = x$ or $a + (a + b) = x$ which can only be solved in one way.

She encountered no difficulty in solving inverted problems of the $a - x = b$ type either. The repetition of such problems revealed only a minor tendency toward likening the statement to some well secured standard in the course of repetition. She could, however, easily correct the mistake and solve the problem.

Here follow some illustrations.

The patient is given the problem: *A boy had 24 apples; he gave some of them to another boy. Thereafter he had seven apples left. How many apples did he give to the other boy?*

The patient repeats the problem: "The boy had 24 apples. He gave seven apples to another boy. How many apples were left?" *You repeated it incorrectly. Repeat it more attentively.* (The problem is read anew.) "Oh, yes, I repeated it incorrectly." (The statement is repeated correctly.) *Now solve the problem.* "Well, that's easy — he gave away 17 apples."

A trend to change the statement according to the habitual stereotype.

This trend is, however, easily overcome by the patient.

Solves the problem at once.

(2) The patient repeats relatively easily and solves comparatively more difficult problems that require two operations and cover the simplest variant of division into parts.

The patient is given the problem: *A housewife used six liters of milk in five days. How much did she use in a month?*

The patient repeats the problem correctly and immediately begins to solve it. "Let's learn how much is used . . . just a minute . . . how many five-days are there in a month — 30 ÷ 5 = 6. Now we can answer the question — 6 × 6 = 36."

The patient encounters no difficulty in drawing up the plan and solving the problem.

It is clear from the record that the patient does not face any serious difficulty in solving this problem. Only the beginning of the impulsive question "Let's learn how much is used . . ." which is quickly inhibited and replaced by the correct phrasing of the intermediate question and the planned solution of the problem, testifies to certain peculiarities in her thinking.

It is noteworthy that this tendency towards impulsive actions may be sometimes more evident, for instance.

The patient is given the problem: *A housewife uses 15 liters of milk in five days. How much will she use in a week?*

The patient begins to solve it at once: "We shall first multiply 15 by five . . . in five days — 35 liters."

The problem is read again. "15 . . . in five days . . . So, we should learn how much in seven days . . . We must first learn how much she uses a day — 15 ÷

The impulsive operation of direct multiplication. The given answer is 35 liters. The traces of the operation 5 × 7 = 35 (i.e., the trend towards another impulsive operation).

5 = 3. Now — how much in a week — 15 + 6 = 21."

The patient quickly draws up the solution plan.

Some operations are performed mentally.

Thus, the solution of this problem (which requires a sequence of operations different from that of the previous problem) revealed some features that we shall later come across, repeatedly describing the patient's intellectual characteristics. She is still perfectly capable of drawing up the solution plan and quickly fulfilling it; but under certain circumstances she may attempt *direct, impulsive* problem solving. These attempts are, however, easily inhibited and give way to the correct solution.

The tendency to impulsive answers may sometimes take even more manifest forms.

Let us analyze how the patient solved a similar problem which, unlike the previous one, covered more elements.

Some collective farmers had 12 hectares of land. They harvested five tons of seed from each one. One-third of seed was sold to the state. How much was left?

The patient immediately says without repeating the statement: "That's clear! 12 × 5 = 60, 1/3 — that's 20 . . . 60 − 20 = 40!" *How much did they sell?* "40. . . , no, one-third of 60. . . one-third. . . how much is this? 30. . ." *What's the answer?* "Thirty."

The quick solution of the problem.

Repeat the problem. "Just a minute . . . that's more difficult for me. Some collective farmers sowed five ha of land . . . How much did they sell to the state from each hectare? Twenty tons . . . One-third. How much did they sell? Thirty . . . The collective farmers sold 30 kilograms of grain. No, not kilograms but . . ."

Perseverance in changing the numbers 1/3 and 30. The scheme of the problem is repeated correctly but with impulsive changes in numbers which point to an unstable system.

The problem is repeated again. "So, some collective farmers sowed 12 kg . . . 12 ha of land . . . (from here on the problem is repeated correctly.) Let's calculate. . . one-third of 60. . . it means 30. . . 30 left." *No, that's wrong.* "One-third of 60 will be 20. . . 20 sold and 40 — left."

At first the patient perserved in repeating the number 30 but the mistake was quickly rectified and the problem was solved correctly.

Thus, we may trace here the impulsively arising substitutions of values which are, however, easily corrected by the patient. She is perfectly able to develop the general solution scheme and determine the operations, which are often performed mentally and are of strictly selective character.

(3) The patient encounters obvious intellectual difficulty whenever the nature of a problem becomes more *complicated* and its correct solution requires *a preliminary orientation of the statement.* Such an orientation should

cover the comparison of its constituent elements and the development of a definite plan (strategy) for problem solving, which will determine all the subsequent operations.

Therefore, it is clear that, in contrast to the cases described in Chapter Two, the preliminary orientation of the problem statement ("the orientation basis for an intellectual operation") can easily be lost here. Therefore, the patient has a tendency to attempt to solve a problem directly without analyzing its statement, identifying the main question and developing the respective solution plan.

It goes without saying, that under the circumstances, the planned process of solving a problem turns into a chain of fragmented operations each of which either answers a specific question, or is altogether withdrawn from the general semantic frame of the problem and is actualized as an isolated numerical operation prompted to the patient by the arithmetic "cues" ("more", "less", "so many times more", "by so many elements less", etc.) in the problem statement. These operations are sometimes so greatly withdrawn from the general solution plan that the patient is unable to say which problem element she is trying to find and which question is being answered by that operation.

It is important that such a direct transition to impulsive attempts to perform certain operations that have nothing to do with the solution plan, can be found even in the case of the correct repetition of a problem statement. Hence, the reproduction of the problem statement (which is not always impaired) cannot in this case be regarded as the basis for preliminary orientation in the problem statement and does not play the regulatory role which it plays in the normally organized intellectual activity.

Let us review some instances, illustrating the patient's behavior while solving difficult problems.

The simplest case is when the patient, having read the statement, omits one of its elements, simplifies it, and answers in the course of problem solving, not the final but an intermediate question of the problem. Thus, a *composite problem* turns into a simple one and the hierarchy of mutually subordinated operations is replaced by a specific direct operation.

Nick had 21 roubles and his brother — three times less. They spent all the money on an aquarium. How much did they pay for it?

The patient begins to solve the problem at once and declares: "One-third — so, it's seven."

The problem is read again. The patient repeats it correctly and says: "Well, it's clear, $21 \div 3 = 7$... So, the aquarium cost seven roubles." The problem is read once more. The patient is asked to follow its semantic parts attentively. "Ah, so! Well, certainly, $21 + 7 = 28$."

The patient performs the first operation of the problem, the other operations get lost, the complex structure of the problem disintegrates.

In this particular case she simply ignored the final question which suggested the addition of two constituent parts; the intellectual operation was simplified but not meaningless. Therefore, the defect was easily compensated for, by simply drawing the patient's attention to the details of the statement while reading it.

The process of solving a slightly more difficult problem was similar in nature.

A son worked 23 workday units, his mother — nine workday units more than the son and his father — 19 workday units less than the son and the mother taken together. How many workday units did each of them work?

The patient repeats the problem and immediately begins to solve it. "The mother earned 32 workday units, now I'll learn how many the father earned: 32 − 19... That's all."

No, you solved the problem incorrectly. Listen attentively. The patient listens to the problem and repeats the statement correctly. "How many workday units did the father work? The father got 19 workday units less than the son and the mother taken together. So, it's 32 − 19. These are all the father's workday units."

The statement is repeated. This time emphasis is laid on the words "19 workday units less than the son and the mother taken together." The problem is solved correctly.

It is important to note that even when the patient correctly repeated the part of the statement that required additional operations implied in the statement (the addition of two elements) she continued to perform nothing but the specific operations overtly formulated in the statement. Hence, her intellectual activity was regulated *only by the overtly formulated statement of a problem.* We had to accentuate the part of the statement which required the additional operation to make it part of the patient's intellectual activity.

(4) Hitherto the patient's intellectual operations, which were simplified due to the withdrawal of one of the statement components, remained quite meaningful and did not differ much from normal intellectual activity. But with the further complication of problem statements, the tendency to replace the preliminary development of a solution strategy by the impulsive performance of an isolated operation outside the frame of the problem may, however, lead to disintegration of the intellectual operation.

One pen and one ABC book cost 37 cents, two pens and one ABC book cost 49 cents. How much does one pen and one ABC book cost separately?

The patient repeats the statement correctly and says: "Let's look at the second boy. We shall determine how much more he paid — 49 − 37 = 12. This is the cost of one pen. Now 12 ÷ 2 = 6. Now we shall learn how much one ABC book costs... 49 − 12... if each costs six... then two will give 12... 37... (consults the statement) "There is something wrong here"	The patient begins to solve the problem correctly but is directly affected by the fragment "2 pens" and halves the result. The patient impulsively subtracts the bigger value from the quotient and reproduces the already performed operation.

The problem is repeated in parts and explained to the patient. She correctly repeats the statement and solves the problem.

It is typical of this case that the patient correctly draws up the program of the operations; the program is, however, so unstable that it easily disintegrates under the influence of the partially perceived statement element (two pens) and the subsequent operations are no longer performed in keeping with the original plan.

(5) We have described the difficulty faced by the patient when the problem gets more complicated and, in particular, when the solution algorithm covers several questions. She encountered even greater difficulty whenever she was offered problems that required the distinguishing of an intermediate link with the subsequent inverted operation. But even these problems were relatively well-solved by the patient. For instance,

A son is five-years old. In 15 years his father will be three times older than the son. How old is the father now?

The patient repeats the problem correctly. *Solve the problem.* In 15 years... The father will be three times older than the son . . . I don't know how to solveit... If I knew I would solve it." *What should we learn?* How old the father will be in 15 years... This should be learned first—or... "*Cannot we learn this at once?* The patient repeats the statement. "I believe he is three times older now too." *Why?* "According to the statement... This condition persists. If he is originally three times older then he is presently three times older too, isn't he?"

The patient is given an example, illustrating that with age the ratio changes. "Yes, most probably, that's wrong." *How old is the son now?* "He is five." *In 15 years the father will be three times older.* How old will the father be in 15 years." *And how can we learn that?* That's the point... *How old will the son be in 15 years? And what about the father?* "He will be sixty."*Now solve the problem.* "The son is five-years old. His father is three times older... the father will be three times older in 15 years... How old was the father originally? 60 − 15 = 45."

The question is worded impulsively, without taking into account if it is possible to answer it.

The ratio "3 times older" is analyzed outside the statement context.

The patient again words the question without taking into account if it is possible to answer it. She does not formulate the intermediate unknown. She simplifies several elements while repeating the statement. The patient can solve the problem after it has been correctly formulated.

The presentation of a similar problem for control purposes shows that the patient easily transfers the mastered algorithm of problem solving. Hence, the complication of a problem and the need to formulate a question about an unknown intermediate value, may lead to some difficulties, as a result of attempts to solve the problem without any preliminary orientation of the statement. These difficulties are, however, successfully eliminated with the investigator's help, and the patient masters the solution algorithm for the given problem.

(6) Thus far we have dealt with relatively complicated problems which were, however, fairly easy for the patient. She encounters much more serious difficulty when dealing with so-called standard problems which cannot be solved by usual methods and require the use of special auxiliary techniques.

Our research findings show that the patient repeatedly tried to solve such problems by a series of impulsive operations, and that these impulsive attempts hindered the identification of the correct algorithm.

We shall first present the results of a study conducted a few days before the operation when the focal symptoms were aggravated by diffuse brain disorders, and then present the findings of the study conducted with the same patient several months after the operation, when these additional impairments had disappeared.

The patient who has already solved the problem: *"There were 18 books on two shelves, but on one of them there were twice as many books as on the other"* gets the following problem:

There were 24 kg of apples in two baskets but there were three times more in one of them. How many kg of apples were there in each basket?

The patient repeats the problem correctly. *Now solve it.* She immediately answers: "Three and eight in the other..." *Why do you think so?* "I multiplied eight by three . . . There is such a solution method. *Which method?* "That's the point. My memory is no longer good." (The problem is repeated again.) "Yes, that's it. I'm solving it this way: there are three kg in one basket and three times more... nine kg... No, that won't do!" *But you've solved similar problems!* "Then I must find which number is three times more than three... No, I can't do that." *Is this a problem in parts?* "Yes, it is. It's a problem in parts but I can't solve it."

The patient performs the direct operation of tripartition.

The operation "three times more" is replaced with "three kg more."

The patient is once again given the previous problem: *There were 18 books on two shelves but on one of them there were twice as many books as on the other. How many books were there on each shelf?*

"There is one part and two parts here. All in all there are three parts. I think it must be solved this way. There are two books on one shelf and the rest of the books are on the other shelf." *But this is a problem in parts.* "Ah, then 18 ÷ 2 − 6 on one shelf and 12 on the other. So simple!"

Sliding to the operations with the books.

Successful solution of the problem.

In order to check whether the patient can tell the algorithm of this problem from that of another problem which resembles it on face value alone she is offered the following problem: *There are 24 books on two shelves but on one of them there are two less books. How many books are there on each shelf?*

The patient repeats the problem correctly. *Solve the problem.* "Well, I'll solve it but I'll think for a long time. Well... 22 plus 2 gives 24. Two books less than on the other shelf. No, that's wrong." *And how will you solve it?* "24 ÷ 2 = 12... Two should be subtracted... Ten and 14." *Is that correct?* "I don't know if that is right or wrong... I feel so bad to be unable to solve such a simple problem!"

The patient performs the first operation directly prompted by the statement and stops there.

The patient is prompted to the solution method and she solves the next similar problem correctly. But when she returns to the previously solved problem she once again encounters the same difficulty.

The patient gets the problem: *There are 36 pupils in two forms but in one of them there are twice as many pupils as in the other. How many pupils are there in each form?*

The patient repeats the problem well. *Solve it, please.* "I can't... I have been lying and thinking... how to solve it... I don't know what to begin with." *You have already solved such problems.* "I know I have... but I can't. There are 36 pupils... in two forms. In one of them there are twice as many... Well, how shall I solve it? 36 ÷ 2 = 18 But what shall I do to have twice as many in the other form?!" (she refuses to go on with the solution.)

Against a background of impaired consciousness and inadequate reflection of the environment the problem is not solved at all. The patient refuses to solve the problem.

Therefore, the experiment conducted at a time when the impairments caused by the local lesion of the left frontal lobe were accompanied by the additional diffuse brain (cerebral edema) disorders testify to the marked disturbance of intellectual activity. The patient generally performs only some impulsive operations and does not progress to the development of a general solution scheme. She can grasp the method for problem solving only having received the respective assistance from the experimenter. She encounters difficulty, however, in transferring this to the subsequent problems and continues to limit herself to fragmented operations that arise. Despite all our efforts she did not progress to the planning stage of problem solving.

The repetition of the same experiment a month after the operation graphically illustrated the features of intellectual changes caused by the focal lesion of the basal-frontal brain areas.

As the diffuse brain components manifest in the patient's inactivity had been withdrawn, we clearly could see the impairments of intellectual processes typical of this group of patients that manifested themselves as impulsive operations.

The patient is again offered the problem: *There were 21 books on two shelves but they were unequally distributed: on one of the shelves there were twice as many books as on the other. How many books were there on each shelf?*

The patient repeats the problem correctly. "I don't know how to solve it. If we could halve 21... Let's assume I've changed this problem into a problem in parts... Let there be an equal number of books on each shelf. How will that be numerically?...

A tendency to impulsive halving. The patient draws up the correct solution plan.

There was one part on one shelf and 2 parts on the other. How many books will that be? I don't know. If $21 \div 2 = 10.5$ . . . But there are twice as many books there. So... 10.5 books... No, I don't know."

The fragmentary operation $21 \div 2$ is so stable that it blocks the correct solution.

This is a problem in parts. Try to solve it. "I don't know to how many books each part corresponds... I shall after all halve 21 . . . that will be 10.5. . . If there are 10.5 books on one shelf, how many books are there on the other . . . That's the question." *How many parts are here all in all?* "Ah!... three parts! So, $21 \div 3 =$ ... that means seven and 14."

After the significant elements of the statement have been consolidated the patient solves the problem correctly.

This time the patient masters the solution method well and thereafter encounters absolutely no difficulty in solving similar problems.

Hence, the typical impulsiveness that manifested itself during the more acute phase of the disease in solving relatively simple problems persisted with the improvement of her condition only in solving more difficult problems.

(7) Let's consider the series of experiments when the correct identification of the solution algorithm is hindered by the phrasing of a problem statement (we've termed them "conflicting" problems). It goes without saying that such a phrasing of the statement urges the patient to solve the problem directly, instead of developing a complex system of preliminary logical conclusions. This factor aggravates the tendency to substitute impulsive simplified attempts at problem solving for the complex solution method that has been always characteristic of the patient.

Therefore, the patient faced serious difficulty in solving such problems even after the operation when the diffuse brain disorder had completely disappeared.

The patient is given the problem: *A father is 45-years old. He is 19 years older than his son. How old are the two of them together?*

The patient immediately answers:
"$45 + 19 = 64$."

The problem is simplified due to the elimination of

The investigator reads the problem again, stopping at each part of the statement. "Ah! I've become a real dunderhead...45 − 19 = 24. The answer is 24. (She does not notice her mistake.) *And what is 24?* "What is 24... God knows what it is . . . " *Think, please.* (The problem is read once again.) — "Yes... That is the son's age . . . " (She does not begin to perform the next operation.) *Is that all?* "Yes, that's all."

an intermediate element: $a + (a + b)$ is replaced with $a + b$.

The problem is simplified once again: the final question is replaced by an intermediate one, a composite problem turns into a simple one.

The problem is read again. "Oh, yes... how old are the two of them together

...45 + 24 = 69."

The operation is no longer correlated with the question and becomes meaningless.

A day later the patient was once once again given the same problem. The patient repeats it: "The father is 45 years old and he is 19 years younger than his son . . . " (The question is not repeated.)

What shall we do? "We shall add 64 to 45." *What for?* "To learn how old the father is...So, the father is 64 years old." *No, the solution is wrong."* — we should learn how old they are together . . . The father is 64." *But he is 45 according to the statement!* — "Then 64 minus 19 . . . = 49 . . . No, that's wrong... Oh ...45. That's the son's age . . . Too much . . . Now let's learn how old they are together...45 + 64." *No, that's wrong!*

(The problem is read in parts, the investigator asks the patient successive questions for each part and the patient easily solves it.)

A complex method of problem solving is replaced with a fragmented operation. The connection with the final question is lost. A flaw in the problem's wording: "the father was 19 years older than his son" and the idea that the son is 19 years younger than his father. She does not notice that the flawed statement is meaningless. The final question of the problem gets lost. The patient performs the operation $45 + 19$; the question of the problem is lost: she substitutes the already known value from the statement for the final question. We once again see an isolated inverse operation with the patient persevering with the correct numbers.

The given illustration graphically shows that fragmentary operations, substituting for the planned process of problem solving, block the development of the correct program for an intellectual operation and easily lose the connection with the problem contained in the question, turning into operations disconnected from the general semantic idea. Moreover, it proves that the correct program for an intellectual operation, formulated by the patient after repeating the statement, is so unstable, that one day later it does not any longer regulate the solving of the same problem and the process of problem solving is once again replaced by fragmentary attempts that easily lose their connection with the question of the problem.

The tendency to replace the planned process of problem solving by fragmented, often meaningless operations can also be traced when the patient turns out to be able to solve the most difficult problems but quickly loses sight of the goal of the operation while performing it.

A pencil is 16 cm long. The shadow from the pencil is 48 cm longer than the pencil itself. How many times is the shadow longer than the pencil?

The patient immediately answers: "Three times. (The problem is read again.) Oh, now I see:so, it's 48 + 16 = 64." (a pause) — *What does it mean?* — "Well, really. . .what does it mean? . . . (The problem is read for the third time. The patient repeats it correctly.) Then the shadow is three times longer than the pencil." (The problem is read for the fourth time. The patient repeats it correctly.)

An impulsive answer. The operation becomes meaningless.

The return to the impulsive answer.

Solve the problem. "So, I must determine . . . aha! So the pencil is 16 cm and the shadow plus 48 cm, that is, 64 is longer than 16. . . four!" *What is four?* "Well, what is four? I don't know this myself. . . "

After another repetition of the problem the operation is correlated with the statement and the idea is restored.

Hence, even when the operations that are required to solve a problem are performed correctly they easily lose the connection with the final question and become meaningless for the patient. The *tendency to withdraw the operations from the general context of the problem* is very characteristic of the given patient.

Thus, we have described a patient with an intracerebral tumor located near the basal sections of the left frontal lobe, which caused the pathology of the frontal cortical area. We may draw the conclusion that she was, in principle, able to perform intellectual operations while solving arithmetic problems. The patient, however, seldom managed to do it as she tended to omit the necessary preliminary orientation in the statement and to immediately move directly to problem solving.

As a result, she was good at solving simple problems (which did not require any special orientation efforts) but easily substituted impulsive operations prompted by the directly perceived statement fragments, for the solution of complex problems which required special orientation of the statement.

These impulsive, fragmented operations often dominated her activity to such an extent that they blocked any attempts to draw up an adequate program for an intellectual operation. But even when the patient managed to formulate such a program, it turned out to be so unstable that it could easily disintegrate and was often replaced by fragmentary operations. The latter, in turn, easily lost the connection with the statement of the problem and became meaningless.

All these defects in intellectual activity, however, could be eliminated if the patient was asked to read the statement attentively, formulating all the potential questions, and if she was given the successive solution plan she could perform a normal intellectual operation. The methods for overcoming the defects in the course of rehabilitation training will be discussed in detail in Chapter 9.

The disorders of the intellectual processes in the described variant of the frontal syndrome are distinguished by the instability of the complex programs for an intellectual operation and the tendency to replace them by fragmented, impulsive operations. These impairments, however, can be compensated for.

Disorders in Problem Solving During the Second Stage of the Disease (The Case of Patient B)

The patient was discharged from the hospital in good condition after the operation. But eight months later, the situation changed for the worse: she began suffering from headaches, complained of fatigue; some psychological changes took place. These symptoms grew increasingly worse and in February 1964 — a year after the first operation — she was readmitted to the Institute of Neurosurgery, this time in a considerably altered condition.

The patient was inactive and indifferent to her environment. Her activeness was drastically reduced. Her movements were characterized by the tendency to "freeze."

The eyeground had marked papilledema. She suffered from decreased visual acuity. She had paresis of upward gaze. Her sense of smell remained intact. She developed the central paresis of the right facial nerve, weakness in the right hand, the dominance of the right tone by the extrapyramidal type, overactive reflexes on the right — occasional palmomental and grasp reflexes on the right, the lip reflex. Her sensation was preserved. The EEG showed pathological slow wave activity in both frontal lobes, which were more apparent in the left hemisphere and were most prominent in the left temporal area, against a background of other changes (reduced alpha-rhythm and defuse delta-waves).

All this pointed to the continued growth of the tumor which must have this time penetrated the deep sections of the left frontal area and, probably, the right frontal lobe.

During the *second operation,* performed on February 26, 1964, Dr. Alekseyeva removed a large tumor (anaplastic oligodendroglioma) located deep in the left frontal lobe. The tumor occupied the pole of the left frontal lobe, spreading backward to the pole of the temporal lobe, reached the falx and, probably, spread to the other hemisphere.

The patient's neuropsychological status and specifically her intellectual activity were assessed both before and after the operation.

The general neuropsychological assessment conducted before the operation noted the drastic aggravation of the patient's symptoms.

For example, she could orientate herself in space and time but was inactive, indifferent to her environment, and somewhat euphoric. Furthermore, she lacked spontaneity. She could be easily persuaded to undergo the assessment but, having begun to perform the required operation, she immediately "froze" and we had to continuously encourage her to go on by additional stimuli. She recognized the investigator but her verbal communication was hindered by obvious speech deficits: the patient had to search for words, her speech was full of perseveration errors and she herself noted that her speech "had deteriorated." When asked: *What are you complaining of?* she answered, repeating the question in an echolalic fashion: "What are you complaining of? I'm complaining of nothing." When asked if she suffered from headaches she replied: "My pelvic bones and my right elbow or rather the radius of my right arm are aching all the time... that's why I'm here..."

Is this the only reason for your coming? "Not only... I've had headaches, therefore, I've come." *What else is aching?* "My eyes have been aching... I had a strong pulsation in the head..." *And now?* "And now... now I'm having the feeling that I have no elbow... elbaw... elbaw... lumbar joints..." *And how is your speech now?* "It's getting worse: I want to say something and don't know what it is..." *Why is it so: is it your empty head or does your tongue disobey you?* "It's both, my head is empty and my tongue does not obey me any longer."

An objective assessment gave the following picture. The patient's praxis remained intact but she performed all tests of praxis slowly, with a severe tendency toward persevative in errors, and standing still at a position once adopted. Her optic gnosis is unimpaired. We did not notice any astereognosis. She tended to stick to stereotypes while reproducing the rhythms.

The patient's speech had no phaseal impairments; her word repetition was preserved. While reciting word series she inertly stuck to the established stereotypes. She encountered no difficulty naming objects but her spontaneous speech was characterized by echolalia, difficulties finding adequate words, and perseveration errors. The patient's spontaneous speech was disordered; it was very difficult for her to initiate a phrase on her own. She did not address the interlocutor. When she was asked to name *three red* items she said: "red, terracotta..." and stopped still. When she was asked to name only *red items* she said: "Shoes... bag... gloves." And only when we specified the assignment and asked her to name only *red vegetables* she said: "Carrot, tomato, water melon... may I name water melon because of its inner part?..." When we continued this test and asked her to name *three words beginning with letter "k"* she declared: "Now... now I'm going to name words beginning with 'k'... well... k..." and that was all. After further encouragement she said: "Well, okay, surnames beginning with "k"- Kalashnikov, Kalinin..."

The simplest intellectual operations, such as the identification of simple logical relationships, for instance, those of opposition, were intact and the patient faced no difficulty in performing such tests. But she quickly stopped fulfilling the assignments that required extended operations. For example, when she was requested to count backward from a hundred in sevens she said: "Ninety three... (stopped still)... 80... 80... well... 94."(The intermediate operation 93 – 3 – 4 was partially performed and the patient stopped at the fragment "four", adding it to the persevering number 90, with the resulting contamination "94.")

The patient could retell simple stories: her retelling was, however, quickly hindered by inert stereotypes. For example, when retelling the story "The Jackdaw and the Pigeons" the patient said: "The jackdaw heard that the pigeons were well fed... She flew into the house... where the pigeons were fed and cried in the jackdaw fashion... But this cry was unexpected for the pigeons and... they drove her out. Then the jackdaw flew to her *pigeons*... not the pigeons but the *jackdaws*... and the pigeons also drove her out... not the pigeons but the jackdaws... "

What did they drive her out for? "Well, because she is a swindler. *What does it mean?* "Because she exchanged work for idleness... Well, generally speaking... she exchanged a jackdaw for a trifle[3]... and she exchanged the trifle... well, she exchanged a jackdaw for a jackdaw... "

Despite numerous repetitions which hindered the retelling of the story, the patient grasped its general idea. When asked about the idea of the story, she said: "To adjust oneself."

It is noteworthy that the patient who failed to retell a more complicated story unaided, could easily retell it answering separate questions.

All this points to the evident syndrome of dynamic aphasia which developed against a background of decreased spontaneity and pathological inertness, and measurably hindered the normal operations of verbalized thinking.

Shortly before the second operation, we reassessed the patient's problem solving process which had been first studied during her previous stay at the Institute of Neurosurgery. The assessment showed an aggravation of the syndrome of intellectual disorders caused by impulsive, fragmented answers. The aggravation resulted from the growing general inactivity and the pathological inertness of established stereotypes, as well as the symptoms of dynamic aphasia, described above. All of the problems were solved very slowly; each question required additional encouragement; the patient stopped still after each answer. She sometimes persevered in the phrasing suggested to her or inertly repeated the number she had already stated. She repeated in an echolalic fashion one or another fragment of the statement many times and often omitted the question of the problem while repeating it. She neither carried out the preliminary analysis of the statement, nor distinguished its significant elements. The patient often substituted fragments of operations for the true process of problem solving. Such fragmented operations were easily disconnected with the main statement and became meaningless.

To provide the correct actualization of the patient's intellectual operations, the experimenter had to split up the statement into successive elements because that was the only way the patient could perform the required operation.

The patient is given the problem: *47 pine-boards and 35 lime-boards were brought to the workshop; 59 boards were used to make benches. How many boards were left?*

The patient repeats the problem correctly. She is asked to formulate the solution plan. (a pause) *What should we learn?* "How many boards were left." *How can you learn this?* "I would add... 47... No, I would do the following: 47... minus... no."

The patient immediately begins to perform the fragmented operation of subtraction without having performed the preliminary operation of addition.

Repeat the problem. "47 boards were brought... to the local charity, let's assume that... some lime-boards were also brought there... After they had been used to make boards 47 of them were left... no... not 47... but several were left..."

Verbal intrusions (local charity) and perseveration errors (to make boards). Number 47 is retained and easily moves to the place of the difference. The resultant sum is also retained and begins to move to the place of the subtrahend.

The problem is read again and the patient is asked to begin to solve it at once, without repeating the statement. "45 + 35... from the resultant number of boards..." *What should we do?* "We subtract." *What shall we subtract?* "The resultant sum... 47 + 35... We must subtract it..." *Did you solve the problem?* "Nothing to solve here, the problem is so simple."

Thus, the patient retains the correct solution scheme for a simple problem. But the pathological inertness of the established stereotypes leads to the disintegration of the scheme in the course of problem solving. The resultant element is maintained and makes the correct solution impracticable.

The above pathological inertness of the established verbal stereotypes turns out to be so blatant that the patient fails to solve problems on her own. Only when she gets an opportunity to control the process of problem solving, continuously comparing it with the written text of the problem, does she manage to solve it.

The same problem is read again. The patient repeats it well. *Tell me how you are going to solve it. What will you learn by the first and the second questions?* "By the first question I shall learn... how many boards were used... and the second question... and the second question... I don't know how to solve it."

Attempting to solve the problem, the patient substitutes in an echolalic fashion the statement of the problem for its question. She fails to distinguish the questions.

The problem is presented to her in writing. The patient is requested to follow the statement and solve the problem.

Using reading to control the statement, the patient solves the problem correctly but transforms the

"...37...no, 35 + 47...this is the total number of the brought boards...that will make...82...and now in order to subtract...in order to subtract... 82–59. In this way I'll learn how many boards were left for benches." *And what is 59?* "Well, certainly, 59 boards were used to make benches." *And what did you learn?* "I learned how many boards were left."

meaning of the answer obtained, repeating as echolalia the last part of the statement ("59 boards were used to make benches"). She could formulate the answer correctly only after some additional questions.

Whereas the solution of simple problems ([a + b] − c = x) remains easy for the patient, she encounters much greater difficulty in solving more difficult problems that require the preliminary division or the computation of the constituent parts. The process of solving such problems is easily affected by the repetition of preceding operations or is replaced with fragmentary operations removed from the general problem scheme. The patient can solve such problems only if she can continuously verify her problem solving by comparing it with the statement of a problem.

The patient gets the problem: *A housewife uses six liters of kerosine in five days. How much kerosine does she use in 30 days?*

The patient repeats the problem: "A housewife uses in six liters...no, but I'm incorrectly...The housewife uses in five liters...no, in five days... six...liters. How much was left...for 30 the days."

The repetition of the statement is complicated by the repetition of denomination (liters) and repetition of the previous problem.

Tell me your solution plan. "6 × 30..." *What will you learn by it?* "How much money she will spend." *What were you to learn?* "How much was left." *Repeat the problem once again.* "A housewife uses six liters...in six days. How much will be left...How much she...be left...no, 5 liters + 30 = 35."

An impulsive operation of multiplication. The question of the problem is transformed by the repetition of the previous problem.

The repetition of the number (6) during the repetition of the statement. The solution is replaced by the impulsive operation of addition (the repetition of the preceding problem).

The patient is allowed to read the written text of the statement and solve the problem, continuously comparing the process of problem solving with the text. Under these conditions she solves the problem correctly.

Hence, when the patient has to solve the problem mentally, reproducing the instruction from memory, we observe an apparent disintegration of her mental operations. Under these conditions the inertness of the established stereotypes is especially conspicuous and organized intellectual activity is very easily replaced by impulsive fragmentary operations.

It goes without saying, that all these symptoms become even more apparent whenever the patient begins to solve more difficult, standard problems.

The patient is given the problem: *There were 18 books on two shelves but they were unequally distributed: on one of the shelves there were twice as many books as on the other. How many books were there on each shelf?*

The patient repeats the statement: "On two shelves there were twice as many books . . . How many shelves with two books were there . . . No, no, it does not matter . . . " The statement is read again. "In the street . . . oh, no . . . on two shelves there were 18 books (she is distracted and begins to sing something.) But I do not know how to solve this book . . . "

The problem is repeated with the contamination of two parts of the statement. Later the statement is completely altered.

The problem is given to her in writing. The patient reads each part of the statement separately and repeats it, answering questions. But having repeated the statement she refuses to solve the problem and says: "No, this problem cannot be solved, it's inconvenient to solve it."

The patient fails to move unaided to the performance of the intermediate question and says that it is impossible to solve the problem within the frame of the given statement.

Even after she has been prompted that this is a problem in parts she does not begin to solve it. The experiment is suspended and resumed the next day.

The problem is read aloud and the patient is asked to repeat it. She says: "On two shelves . . . there were 18 books. On the second of them there were twice as many than on the third . . . It's too much for me . . . "

The repetition is characterized first by the perseverating number "two shelves" and then by the continuation of the natural series.

The patient correctly repeats the statement after it has been split into parts and the beginning of each phrase has been prompted to her

The patient restores the logic of the problem solving with the help of separate questions.

When asked how she is going to solve the problem the patient reproduces the cue given to her the day before and says: "This is a problem in parts." *How many parts will there be?* "Three." *How will you learn the value of each part?* $18 \div 3 = 6$ "*What must you learn?* "There were twice as many books on the shelf . . . so six by two . . . is that correct — shall I multiply? . . ." (stands still) *Well, how much will be six by two?* "Three! I've already forgotten whether I must divide or multiply. . . . "

But even a short pause is enough to lead to the replacement of the correct operation of multiplication by the meaningless operation of division.

Thereafter the patient refuses to solve the problem and when asked "How many books account for three parts?" she answers: "Thirty-five."

She cannot solve the problem even with the help of the written text. The patient says: "I said 35 to you then. . . 35 is out of place here. . ." but she does not begin to solve the problem.

Hence, even having grasped a solution scheme for a problem (if its statement is split into isolated parts) the patient is unable to control the process of problem solving by an adequate solution plan. The operations get easily removed from the statement. The patient fails to take the correct option out of the two possible meanings of the operation "six by two" (division versus multiplication). Therefore, the overall process of problem solving becomes inaccessible to her.

Similar impairments were also observed when she solved other problems. For example, the patient could correctly reproduce the statement after separate questions, but was unable to integrate isolated fragments into a single program of operations on her own. She was absolutely helpless unless the operations required to solve the problem were unambiguously worded in the statement.

The tests were suspended as a result of her aggravated condition and resumed after the operation.

Neuropsychological assessment ten days after her surgery noted the following clinical picture: the patient is more active than before surgery. Her movements are less retarded, the tendency to "freeze" is less evident. She is oriented in space and time and is more willing to perform the assignments. The tendency to echopraxia still persists. For example, she often repeats the required movement in a mirror-like manner, as she omits the operation of recoding which is necessary to reproduce it correctly. The rhythms are reproduced faster but here too she often gravitates toward the inert repetition of the once-established stereotype.

Her speech still contains the elements of echolalia, the repetition of word series is still often impaired by perseverations. The patient can count well, and is able to do simple sums but whenever she deals with problems consisting of a series of elements, she is incapable of performing the required operations. Severe perseverations are still conspicuous in all her intellectual tests. Hence, her post-surgical condition was characterized by an apparent persisting frontal syndrome with symptoms of repetition of inert stereotypes, both in the motor sphere and in speech.

Such was the patient's condition when we resumed our experiments.

On the 10th day after surgery (6.03.1964) the patient is given the following problem: *There were 18 books on two shelves; on one of them there were two more books than on the other. How many books were there on each shelf?*

The patient repeats the statement correctly but words the final question in an echolalic fashion:

The question is again replaced by the echolalia

"How many books were there on the shelves?" When asked how many books were actually on the shelves she replies: "Eighteen"and gives the following answer to the question *What is unknown then?*: "How many books were there on the third shelf?"

of the statement elements. Later the question is replaced with the inert continuation of the numeric sequence.

When the problem is repeated for the second time the patient reproduces the statement correctly.

Solve the problem. " . . . We shall divide 18 by two . . . and then add 18 to . . . 18 . . . and then . . . we shall add to 18÷2=9 we'll add two . . ." *So, what will the answer be?* "So, on one shelf there were nine, on the other — nine plus two . . . so, 11. That's right." *No, that's wrong.* The patient refuses to go on solving the problem.

The complex program of problem solving is replaced by two fragmentary operations.

The second fragmentary operation (18+18) perseveringly repeats the statement of a problem in parts.

The patient is prompted how to solve the problem: *Let's assume that the two extra books are taken away.* "Then 18÷2=16." *What shall we do next?* "What should be done next? I think 18÷2=9."

The investigator prompts the method for solving the problem in order to check later if she has mastered it.

The prompted solution does not form the necessary program and the patient continues to reproduce the previous stereotype of the operation.

Three days later (9.03.64) the patient is given the same problem. She stereo-repeats it well. When requested to solve the problem she says: "Well, yes, 18÷2=9. . . No, what am I doing this for! So, there are two extra books. . . "After these words she reproduces the solution method mastered the day before.

At first she tends to reproduce the same typed operation but then corrects her mistake and applies the mastered solution method.

The patient easily solved a similar problem differing from this one only in numbers, which was offered to her immediately afterwards. But when asked to formulate its solution plan, the patient inertly reproduced the solution scheme for the problem which had been given to her before the operation and said: "We should learn how much each part is." She corrected the mistake only after the respective question. A similar tendency to alter the statement by simplifying the program implied in it, repeating some of its fragments and substituting fragmentary operations disconnected with the statement, for the discursive process of problem solving is equally evident in solving problems that involve the complex system of intermediate operations.

Twenty days after the operation, the patient was given the following problem: *A boy bought one pen and one ABC book and paid 37 cents for them. Another boy bought two pens and one ABC book and paid 49 cents for them. How much do one pen and one ABC book cost separately?*

The patient repeats the problem persevering its parts: "The boy bought a pen and an ABC book and paid 37 cents, another boy bought a pen and an ABC book and paid 49 cents for them. How much do a pen and an ABC book cost separately?" After the problem has been analyzed the patient repeats the statement correctly.

She repeats "a pen and an ABC book" twice. She does not grasp the semantic structure of the problem which lays the basis for its solution.

Solve the problem. "Well, an ABC book and . . . two pens . . . and one pen . . . and we should learn how much all this costs, is that right?" (a long pause) *Solve the problem.* "I believe, it's 37 + 49 . . ." *What do you learn by it?* "How much money did the boys spend." *Have you answered the question of the problem?* "I think, I have."

And what were you to learn? "How much money the boys spent . . . But I didn't solve the problem this way . . . " *What should you learn then?* "How much did a pen and an ABC book cost to one boy and how much did an ABC book and two pens cost to the other boy."

The enumeration of two sets of purchases leads to the question on the sum total of the purchase that simplifies the question. The simplification of the problem leads to the replacement of the recently reproduced question. The patient has the feeling that the questions disagree but when she attempts to formulate the question she replaces it by the echolalic repetition of the statement elements.

The continued tests show that the patient cannot solve the problem on her own.

These examples illustrate the severe intellectual disorders which were seen during the second period of patient B's disease.

The continual growth of the intracerebral tumor which by then occupied the entire pole of the left frontal lobe, retained its primarily medial-basal location, spread to the subcortical motor centers and partially to the systems of the anterior-temporal area, led to the development of the syndrome which featured, besides the earlier described disorders, some newly-emergent symptoms.

In the first period of the disease the syndrome centered on an impaired orientation basis for intellectual activity and the impulsive fragmentary attempts at problem solving. In the second period it was measurably aggravated by the patient's marked inactivity, which resulted in the tendency to ignore the question of the problem and to replace it with the echolalia of some statement fragment. Her condition was characterized by the conspicuous tendency to substitute the inert stereotypes of the previously used solution methods for the necessary processes of reasoning. Furthermore, there developed some signs of severe verbal perseverations which often resulted in the alteration of the problem's numerical structure and sometimes that of its semantic structure.

As a result of all these factors the patient could not solve any difficult problems that required intermediate operations on her own. When given such problems, the patient either refused to solve them altogether or simplified the problem statement and replaced the proper process of problem solving by fragmentary stereotypes.

The specific feature of this patient was, however, that even in the worst period of her disease she remembered the statement of the given problem and did not slide into side-associations lacking selectivity.

Conclusions

This chapter was devoted to a detailed description of the problem solving disorders observed in a patient with an intracerebral tumor located in the basal sections of the left frontal lobe.

The patient faced no difficulty either in understanding logico-grammatical relationships or performing arithmetical operations — the defects which are typical of patients with an injury to the parietal-occipital sections of the left hemisphere. Although these components of her intellectual activity were intact she was practically unable to solve arithmetic problems.

Serious defects revealed themselves, above all, in her orientation-searching activity: active orientation of the statement of the problem was replaced by impulsively emerging fragmentary operations. As a result, the process of problem solving, deprived of this essential element, was no longer a successive chain of logical reasoning. It turned instead into a series of disconnected operations. The control, that is, the comparison of the obtained results with the original data, was also disturbed.

Such a disorder of an active purposeful activity, controlled by the problem statement, was the main feature of the patient's intellectual characteristics.

In the first period of our observations, when the tumor affected the basal areas of the left frontal lobe and caused the disinhibition of behavior typical of the basal-frontal syndrome, these defects revealed themselves in the difficulty faced by the patient in retaining the proper preliminary orientation in the problem statement. Furthermore, they were manifested in her easy sliding into impulsive fragmentary operations when dealing with complicated problem statements. At this time the defects of problem solving were relatively inconspicuous and could be easily compensated for by the investigator's organizing instructions.

In the subsequent period, as a result of the growth of the tumor, the impairments of problem solving became more marked and became more severe. The patient began to exhibit signs of inactivity (typical of frontal dynamic aphasia) and the pathological inertness of established stereotypes. In this period the loss of the correct program in solving difficult problems became increasingly evident.

But even in such an aggravated condition the externalization of all the statement elements and the external organization of the patient's activity resulted (with certain help from the investigator) in compensation for the above defects (see Chapter 9).

These symptoms made up a picture of problem solving disorders observed in the lesion of the basal sections of the left frontal lobe with the ensuing development of the basal-frontal syndrome.

---

[1]See: The Frontal Lobes and the Regulation of Psychic Processes, ed. by Luria and Khomskaya, Moscow, 1966.

[2]The methods for overcoming the defects in the course of rehabilitation training will be discussed in detail in chapter 9.

[3]In Russian the words "byez*dyelye*" (idleness) and "byez*dyel'itsa*" (trifle) are single-rooted words. Translator's note.

Chapter 4

IMPAIRMENTS OF PROBLEM SOLVING WITH POSTERIOR-FRONTAL SYNDROME

In the previous chapter we discussed the intellectual disturbances typical of patients with injuries spreading mainly to the basal sections of the frontal lobes. In this chapter we shall deal with another group of patients in order to determine other factors which measurably impair the actualization of intellectual processes following lesions of the frontal lobes. These factors are general *inactivity and the pathological inertness of established stereotypes.* Such intellectual impairments are particularly typical of patients suffering from the frontal syndrome.

These factors, which are manifested in any massive lesions of the frontal lobes, are especially conspicuous following lesions of the *brain's posterior-frontal areas.* In these cases, general inactivity and pathological inertness are often revealed against a background of the relatively intact structure of intellectual activity as a whole.

We shall first discuss lesions of the posterior-frontal areas of the left hemisphere. They cause the pathological inertness of established stereotypes, which is especially evident in the *verbal sphere.* It reveals itself, in the first place, in the performance of individual intellectual operations without disturbing the overall structure of intellectual activity. Only then shall we discuss more massive lesions when the pathological inertness of higher mental functions is more conspicuous and goes beyond the difficulty in performing individual intellectual operations. In these cases we observe severe disorders of the entire structure of intellectual activity.

Such sequencing will help us distinguish the factors underlying intellectual disturbances in the lesions of the frontal lobes, and analyze the most complex components of the frontal syndrome.

Patient B: General Information

The patient (case history No. 27,715) is 52 years old, a scientific worker, and a soil scientist. He was admitted to the Institute of Neurosurgery in 1959 as a result of a slowly growing tumor in the posterior sections of the left frontal lobe. We periodically assessed his condition for five subsequent years.

In 1944 he began to suffer from epileptic attacks with the twitching of the angles of his mouth. His head was drawn to the right and his right arm and leg involuntarily moved. Then gradually there developed speech deficits. With these complaints the patient was first admitted to the Institute of Neurosurgery in 1949. His objective assessment noted the mild paresis of the right facial and hypoglossal nerves and overactive reflexes to the right which gradually levelled off. The liquid had 0.49% protein, while the cytosis amounted to 4/3. The EEG showed individual epileptic and slow waves, primarily in the left frontal-temporal area, against a background of the normal alpha-rhythm.

The psychological assessment did not reveal any serious disorders, and he was discharged from hospital with the following provisional diagnosis: an inflammatory process localized mainly in the premotor area which affects the opercular area.

The patient resumed his work on the staff of a research institute. His behavior, however, revealed marked changes: he grew passive and unconcerned with many issues which used to interest him. During one of his business trips, he was not given his per diem allowance, nor did he take the necessary steps to get it and actually starved. In 1956 the epileptic attacks resumed. This time they were accompanied by swinging stereotyped motions of his right arm. His general inactivity became increasingly apparent. In 1959, he wrote an article in which he described all the facts completely, but failed to generalize them and he left the article unfinished. His letters also became stereotyped and by mid-1959 he suffered from a marked lack of initiative; he could not say anything coherently. Such was the patient's condition when he was readmitted to the Institute of Neurosurgery in 1959.

The patient remained oriented in space and time but all his reactions were retarded. He would not complain of anything himself, but a careful investigation showed that the patient was aware of his defects and responded emotionally to his failures. He would usually stay in bed and read fiction, saying that reading was his hobby. The patient could not, however, retell what he had read. He was able to reproduce the content of the story only with the help of a series of consecutive cues and questions. His speech was by then full of echolalia. All his attempts at extended speech ended in failure.

The neurological assessment did not reveal any sensitivity disorders but noted the paresis of the right facial nerve, the reduced force and hypertension in his right hand, the tremor to the right, and pathological reflexes which were especially marked to the right and the lip reflex.

The liquid had 0.60%. protein while the cytosis amounted to one-third. The EEG showed pathological slow waves which were dominant in the left frontal area. The arteriogram showed the dextropositioned knee of the anterior cerebral artery.

The psychological assessment noted severe aspontaneity. His praxis was impaired by the dynamic type and his movements were no longer automatized. The patient exhibited in some tests signs of mirror-like reproduction of the given positions and apparent perseveration errors caused by reference to previously performed movements. We noted neither language comprehension impairments nor disordered articulations, but the repetition of a series of words or phrases showed severe carry-over problems. The fulfillment of any tasks involving the motor components revealed the severe inertness of his nervous processes. As a result of pathological inertness, the patient could neither draw a series of figures, nor perform the operations of writing and calculation. Pathological inertness in these cases was so great that he failed to switch from one operation to another, and perseveringly reproduced the element of the operation which he had already performed. The persevatory agraphia and acalculia, characteristic of the patient at this time, were described by one of the authors in a different work (see: Luria, *The Disorders of Writing and Calculation in Case of Inert Higher Nervous Processes,* 1960 [unpublished]).

On November 18, 1959 the patient was operated on by Prof. Egorov, who found a tumor well-separated from the brain tissue in the parasagittal sections of the left posterior-frontal area. The tumor was adjacent to the dura mater encephali and penetrated deep into the left frontal lobe in the posterior areas, where it infiltrated the brain matter, having no definite borders with it. The tumor was very large. It was adjacent to a large falx section and, probably, spread over to the medial sections of the right hemisphere. The tumor was removed subtotally, its remains were coagulated at the lower edge of the falx. Histologically, the tumor was an oligodendroglioma.

After the operation, the patient's condition drastically improved. The severe aspontaneity and pathological inertness in the motor and verbal spheres gradually disappeared. He found it much easier to answer questions and there were no echolalias in his replies. The patient could coherently retell a story and sometimes entered into conversation on his own initiative and maintain it. The manifestations of persevatory agraphia and acalculia disappeared completely and he could resume his work at the research institute. He continued to write his doctor's dissertation, working, however, part-time.

This condition of compensation lasted for about two years, but at the end of 1962, the symptoms of inactivity began to grow increasingly manifest once again. He was less active in communicating with other people, ceased playing the violoncello, pleading insignificant and easily obviated circumstances. The patient stopped going to work, and soon left his job altogether. His epileptic attacks resumed, accompanied by the loss of consciousness.

At the end of 1962 he was readmitted to the Institute of Neurosurgery, where we observed the increasingly evident symptoms of aspontaneity, inactivity, and motor speech disorders. His behavior was perfectly oriented and adequate, but the signs of inactivity were quite marked and gradually grew increasingly conspicuous.

His praxis remained sufficiently intact. We noted neither the apraxia of position nor spatial apraxia, but the respective tests revealed manifestations of mirror-like reproduction and echopraxia, which he easily corrected when his attention was drawn to them. The assessment of his dynamic praxis revealed bradykinesia. He stopped at each element of the performed assignment (each motor element required an additional stimulus). We also noted severe pathological inertness. Towards the end of 1963 it was already much more difficult for him to switch from one movement to another. When he attempted to perform the test of pressing an item successively with the first, second and fifth fingers, he replaced the required pressure sequencing, either with pressing the item consecutively with each finger (the first, second, third, fourth and fifth) or by the stereotyped alteration of the same fingers (the first, the second; the first, the second). Having mastered this motor stereotype after special training, the patient failed to move to the opposite task and continued to inertly stick to the previous stereotype, instead of performing the movements required by the instruction (the fifth — the first finger; the fifth — the second finger).

He faced similar difficulty in the switch-over test in writing: he froze on the very first element of the drawing and inertly reproduced it instead of moving to the next element. Similar manifestations of pathological inertness were noted in the reproduction of motor rhythms. Beating the rhythm "two stokes," the patient easily slid to the "three strokes" rhythm, inertly persevered in it and did not notice his mistake. Having mastered the rhythm, he could not switch to the opposite rhythm, and perseveringly stuck to the previous rhythmic structure. Attempts to correct the mistake (based on the investigator's instruction) ended in failure, and the patient quickly returned to the previous inert stereotype after a serious of unaccentuated beats. Inactivity and pathological inertness were especially manifest in his speech which became severely impaired towards the end of 1963.

The patient faced no difficulty in language comprehension, he had no articulation defects, and could easily repeat separate words, name the shown items, and answer the questions. But his replies always began with the echolalia of the question. He never gave detailed answers to the questions. It took him a lot of time (so-called latency period) to give a short answer to a question. While repeating the series of sounds, words, and phrases the patient very soon began to substitute perseveration errors for adequate repetition. For instance, after the correct repetition of the sound series "ta-da" and "da-ta" he recited the sound series "ka-ga-ka" as "ka-da-ka." He easily repeated the word series "lyes[forest]-dom[house]-kot[cat]-zvon[peal]" but recited the next word series "cat-forest-house-peal" as "forest-house-cat-peal," persevering in the previous word order, and even as "forest-cat-forest-cat," reproducing inertly the first pair. Even if the experiment lasted for a long time, the inert stereotype neither disappeared nor changed, after several repetitions. When the patient became fatigued, his inert reproduction of the previous elements became even more conspicuous. The manifestations of inertness grew increasingly more apparent, when he attempted to repeat a series consisting of several short phrases. The series "the house is on fire — the moon shines — the broom sweeps" was often reproduced as "the house is on fire — the broom shines — the broom sweeps". He correctly repeated the sentence "There are branching apple-trees in the orchard behind the fence" but persevered into its elements, attempting to repeat the next sentence "A car moves faster than a horse but slower than a plane." He reproduced it as "A car moves behind the fence. . .

behind the fence. . ." Similar manifestations revealed themselves in his attempts to recite an unusual word sequence. For instance, the patient repeated the names of days of the week given in the usual order, but he quickly began to encounter difficulty in repeating them in the reverse sequence. The patient correctly began with "Saturday-Friday-Thursday-Wednesday . . . " but failed to continue it: "Tuesday-Tuesday . . . Monday . . . Sunday . . . on Monday," etc.

All of these manifestations of pathological inertness, which did not reveal themselves in the direct repetition of separate verbal elements, were quite apparent in the reproduction of long verbal series from memory. They were especially conspicuous, if the assignment was completed in the context of sufficiently secured previous stereotypes. These stereotypes measurably disturbed the patient's extended, active speech: it was full of severe contaminations and paraphasias related to the pathological inertness of the established stereotypes, and made extended speech nearly impossible.

All of these manifestations revealed themselves against a background of the relatively intact semantic and logico-grammatical structure of speech: the patient easily grasped sufficiently complex logico-grammatical and logical relationships; he faced no difficulty in understanding such relationships as gender-aspect and aspect-gender, and could think of relationships similar to the given ones if this did not require the extended involvement of verbalized thinking.

Hence, all the observed manifestations of impaired higher mental functions testified to massive *posterior-frontal syndrome*, which was described in some of our earlier publications (see: Luria, 1962, 1963). This syndrome consists of massive disorders in the dynamics of mental processes, that reveal themselves in the pathological inertness of the established stereotypes.

The observed manifestations, however, revealed themselves against a background of the relatively well-preserved structure of the patient's intellectual processes, while the disturbances of the same were also relatively stable.

This was the main distinction of the given case from the syndrome of disordered mental processes caused by an injury to the anterior sections of the frontal area. Therefore, this case was very good for our research purposes.

Impairments in Problem Solving in the Case of Patient B

As noted above, patient B's intellectual activity was characterized by the following distinctive features:

- he could understand and remember a problem well;
- he was generally able to repeat it accurately enough;
- in some cases, when the solution did not require complex intermediate operations, he could draw up the plan, retain the adopted strategy, perform the respective operations, and even solve the problem.

We, as a rule, did not observe in the case of patient B instability of solution strategy or loss of the selectivity of operations that are, as repeatedly noted above, (see also Luria, 1962 pp. 187–253; 1963 pp. 340–380, 430–453) the distinctive features of an extended frontal syndrome. The main distinctive features of our patient's intellectual operations were, however, *an inactivity* in the actualization of his intellectual processes, and *the increased inertness* of established stereotypes (primarily verbal ones), that measurably disturbed the dynamics of his intellectual processes.

The patient's inactivity was relatively inconspicuous in solving simple problems that were well-known from his previous experience. In these cases the sequencing of his intellectual operations, which virtually reproduced the well known operations, was, structurally, close to the norm. But his inactivity grew immediately apparent whenever the patient dealt with problems that were complex or new to him, as he had to actively search for their solution algorithm. In these cases, the patient did not conduct the necessary preliminary orientation-investigation work on the statement of a problem. He either refused to solve it altogether, or easily yielded to the tendency to simplify the statement and substitute common stereotypes for the complicated logical reasoning required by the problem. The chain of necessary logical conclusions often broke and the patient paused, having performed an intermediate operation. He neither compared the performed operations with the final question of the problem, nor realized the incorrectness of his solution. In these cases, the investigator had to resort to *additional encouragement:* he urged the patient to compare the obtained results with the question of the problem, or check whether it was solved correctly. Such additional encouragement was generally enough for the patient to perform the required comparison or correct the obtained answer.

The additional encouragement failed to meet its objective, only in the case of the most difficult (so-called standard or conflicting) problems, as their solution requires the development of a special technique or the elimination of well known stereotypes. When dealing with such problems, the patient continued with the previously adopted simplified solution algorithm and inertly repeated the developed incorrect solution procedure.

It is noteworthy that, similar to other patients with injured frontal lobes,[1] his memory of the main question of a problem was much worse than its "material" components. The final question was sometimes (in the case of difficult composite problems) easily replaced by some intermediate question. We also observed that the patient tended to substitute the repetition of an element explicitly mentioned in the statement, for the final question of the problem. Furthermore, it was much more difficult for him to understand the performed operation, than to perform the respective operations. We often noted that the patient failed to formulate properly the operations which he had correctly developed and performed.

All these defects, which were very marked in patients with massive frontal lesions, were relatively inconspicuous in the case of patient B. Furthermore, they could be corrected easily enough, provided additional stimuli.

The second factor, which measurably disturbed the patient's intellectual processes, was *the increased inertness of established stereotypes.* It was especially evident in his speech. The established stereotype and wording revealed such marked pathological inertness that they blocked any switch-over to a new stereotype or wording. As a result, the patient was often unable to remember and repeat a problem with a number of denominations as he persevered in one of them and, thereby, disordered the system of operations required to solve the problem. The inertness of established stereotypes not infrequently altered the nature of the respective arithmetical operations. For instance, having performed the operation of multiplication, the patient repeated on this operation at the next stage, which required addition.

The pathological inertness of the stereotypes was, in this case, practically inconspicuous in solving simple problems but very apparent when he solved difficult problems, as most of their operations could be performed only if the patient remembered the previous ones. All these manifestations grew especially conspicuous when the patient became fatigued.

As noted above, the pathological inertness was particularly marked in the patient's verbal processes; therefore, *it was much more difficult for him to solve problems aloud than to do it mentally.* The patient differed drastically in this respect, from patients with massive lesions of the frontal lobes and faced difficulties which were similar to those encountered by the aphasics.

Finally, another distinctive feature about this patient was the *ease with which he could correct his mistakes.* The patient remembered the original statement of a problem well, but encountered difficulty in the process of problem solving as a result of inactivity and the pathological inertness of the established stereotypes. Therefore, whenever he faced difficulty in problem solving he compared the obtained results with the original statement and could easily correct his mistakes, provided additional encouragement. Hence, as we shall demonstrate below, the introduction of an adequate *behavior program,* the splitting of operations into consecutive stages or the use of external auxiliary supports, due to which the patient's actions, grew *extended in character,* generally compensated for his defects, and resulted in the correct solution of a problem. The intact system of original intentions and the possibility of compensating for intellectual impairments distinguished this patient, like other patients with injuries to the posterior-frontal sections of the brain, from the patients with the massive prefrontal syndrome whose intellectual disorders will be discussed later.

Now let us review how he solved the increasingly more difficult problems.

(1) The patient repeated and solved without any difficulty the *simplest problems of the type* $a + b - c = x$, for instance: *47 pine-boards and 35 lime-boards were brought to the workshop. How many boards were brought to the workshop altogether?*

He began to face certain difficulty, resulting from verbal carry-overs, only when he was asked to think of a similar problem on his own. The hardest job for him in these cases was to phrase the question of the problem.

Think of a similar problem. "Let it be that...35 willow...and 45...elm rods...were brought...that's	Retarded phrasing of the question. Echolalias and

it." *Is that all?* "No, not all . . . we should ask the question. . . well. . . How many problems will there be. . . no, what will be the answer. . . no, what will be the sum total of the rods. . . "

perseveration errors in the wording of the question.

Think of another similar problem. "I was standing at the gate and saw two chairs . . . and two tables brought. How many chairs and tables will that make?"

Inert phrasing of the statement ("brought").

(2) The patient faced much more serious difficulty when dealing with *composite problems of the a + (a + b) = x-type*. He tended to simplify their algorithm, replacing it by the algorithm of a simple problem, or to substitute an intermediate question for the final question of a problem. Therefore, the solution of a composite problem was replaced by answering the intermediate question. This tendency was especially stable when he was asked to think of similar problems on his own.

The patient was given the problem: *Nick used 27 squared copy-books in a year and five more lined copy-books than squared ones. How many copy-books did he use altogether?*

"This problem can be solved in one operation. . . 25 + 5 = 30." *What shall we learn by it?* "How many copy-books the boy used in a year. I'm adding both: five lined copy-books and 25 squared ones."

The statement is repeated again. The patient repeats the problem and solves it correctly.

Think of a similar problem. "May I use the word" Nick "?. . . Well, then . . . Nick's team scored five goals and their opponents scored six goals. How many goals will that be. . . No,. . . this is a different problem. . . " (He asks to repeat the original problem. After it has been repeated he thinks of a correct problem and solves it well.)

The composite problem is treated as a simple problem which was solved earlier. The numbers are contaminated (he says 25 + 5 instead of 27 + 5). The statement of the composite problem is properly formulated and it is correctly solved at the second presentation. Formulating a similar problem on his own, the patient simplifies its statement. On the second presentation of the problem he makes the necessary corrections.

(3) We observed a similar picture when the patient solved composite problems with more difficult relations of multiplicity [of the following type: a + (a ÷ b) = x].

He could quite easily perform the respective operations but when he attempted to get an insight into the performed procedure, he immediately revealed a tendency to simplify the statement of a problem which had been already observed in the previous tests. His attempts to think of a problem on his own reveal the earlier noted tendency to reproduce established stereotypes inertly.

The patient gets the problem: *George had 21 roubles and his brother three times less. They spent all the money on an aquarium. How much did the aquarium cost?*

The patient repeats the problem correctly and immediately gives the correct answer "Twenty-eight roubles."

The problem is easily solved mentally.

Tell me how you solved it. "How I solved... how I solved..." *So, how did you solve it?* "I learned how much... the entire aquarium cost... 21 + 7... makes 21... no,... 28." *And wherefrom did you get seven?* "And seven cost... the younger brother..." *What did you do to learn it?* "I don't know."

The patient repeats the final question of the problem without distinguishing the intermediate operations performed. He does not have any insight into the performed operation.

Tell me consecutively what you were doing. "Consecutively it will look as follows: I first learned how much the younger brother, 21 roubles... into seven... that is, into three... and got seven roubles. Seven roubles... cost... the younger brother." *Not "cost" but "had."* Yes, that's it... had! I added seven and 21 and got 28 — this is the cost of the aquarium."

Perseverating constructions and numbers.

He develops an insight into the solution algorithm in the course of consecutive retelling.

Think of a similar problem. (The patient keeps silent for a long time, he can't think of a topic.) "Well, for instance, with cookies . . . The elder brother had... three roubles... and the younger — three times less . . . How much did they spend when they bought cookies for this money?"

Inactive search for the topic.

The stereotype of the previous content and numbers.

Solve the problem. "Three roubles... three times less... one rouble. One rouble plus three roubles equals four roubles."

The problem is solved.

Tell me which questions you answered The patient is unable to state the performed operations.

He cannot formulate the operations.

(4) All the previous problems were relatively easy. Even the last one consisted of a series of simple linear operations.

People with brain injuries face much more serous difficulty when they attempt to solve problems with a structurally much more complex algorithm, which requires the preliminary grouping of the statements and relies on an increasingly more complex set of auxiliary operations.

We shall first analyze a complex composite problem, solved in 3 consecutive operations, $x + y + z = a; x + y = b; y + z = c$, in which it is necessary to find the individual values for all the three unknowns.

It will be demonstrated that the planned nature of the problem solving process is retained in this case too, but the patient begins to encounter even greater difficulty due to the inertness of established stereotypes.

The patient is given the problem: *Three boys caught eleven kg of fish. The catch of the first and the second boy amounted to seven kg; that of the second and the third boys — six kg. How many kg of fish did each boy catch?*

The patient is asked to repeat the problem. At first he fails to do it and begins to repeat the problem only after additional encouragement. Like in the previous tests, the weakest element in the repetition is the question of the problem that he replaces by the stereotyped reproduction of one of the statement elements.

"The catch of the boys... no, I've forgotten the problem. (The problem is repeated. A pause.) Three... boys caught nine kg of fish... the second and the third... seven kg and the second and the first... six kg. *"What should we learn?"* How many boys were there... no... how many of the boys... it seems to me, caught..."

He does not remember the combination of the denominations. The question gets lost. The data of the statement is transferred to the question.

The same.

The problem is repeated. "Three boys eleven kg. How many kg of fish did the boys catch altogether?" *Did you repeat the problem correctly?* "No, incorrectly." *Then how should it sound?* "I don't know."

The question of the problem is blocked by the stereotyped repetition of the original statement.

The repetition of this more difficult problem reveals a new phenomenon which was inconspicuous in case of easier problems. Trying to reproduce a more complicated system of logical relationships, the patient loses sight of the *main question of the problem,* inertly replacing it by the original data of the statement. But then, as we are going to demonstrate below, he corrects his mistake and is able to distinguish the correct question.

Tell me your solution plan. How are you going to solve the problem. "I'll first learn how many boys... How many boys were caught... No, no... How many boys . . . I shall first learn how many boys were caught no, not this way... How many kg of fish did one boy catch?" *Which one?* "The third and then the second." *And then?* "The first." *And then?* "And then I'll sum up and learn the amount of fish. No, the amount of fish is, of course, known."

The inert reproduction of the stereotype blocks the correct phrasing of the question.

The selective question is replaced by the backward enumeration of the whole statement followed by the inert operation of addition.

The quoted extract from the protocol demonstrates that inert stereotypes not only prevent the patient from distinguishing the proper sequence of questions, but may also result in an inadequate stereotyped solution algorithm (first to distinguish the components, then to sum them up). It is characteristic of this patient that he can correct the inadequate stereotype. It will be shown below that, in contrast to him, people with massive injuries to the frontal lobes are no longer able to do this.

Therefore, the inert stereotypes, that crop up at the presentation of more difficult problems, are a serous obstacle in problem solving. They replace the question of the problem, and hinder the performance of the necessary operations and the assessment of the obtained results. All these factors reveal themselves when the problem is solved mentally after its statement has been memorized and, as shown below (see Chapter 9), they may be eliminated with the help of visual supports which do away with the need to depend exclusively on mnemic processes.

It is noteworthy that the patient sometimes performs some important operations mentally and in this case the perseveration errors, manifest in his external speech, are less detrimental to the performance of the required operations. Hence, we may assume that many impairments, observed in the patient's intellectual activity, are caused by the pathological inertness of his speech. Therefore, the elimination of speech deficits measurably improves his intellectual activity.

(5) Thus far we have dealt with the problems whose algorithm required either linear operations or the consecutive grouping of operations and the assessment of the obtained values through their comparison with the original statement.

From now on, we shall deal with a more complicated algorithm of intellectual operations. We are going to discuss arithmetic problems whose statements do not explicitly list the required operations. In this case it is essential first to derive an intermediate element, which is not overtly formulated in the statement, and then perform the "inverted" operation with it.

Such problems may be illustrated by the following example: *A son is five-years old. In 15 years his father will be three times older than the son. How old is the father now?*

The solution algorithm of this problem is as follows:

$$(a + b)\,n = y;\; y - b = x$$

The patient encountered even greater difficulty in solving such problems, which was also caused by the pathological inertness of established stereotypes. In this case they make it much more difficult for him to solve the problem, as the patient operates inertly with the data of the statement, fails to identify adequate solution methods, and replaces them by fragmentary operations.

The patient is given the above problem which he had already solved earlier.

The patient tries to repeat it. "The son is five-years old, we should learn how old his father will be in 15 years."	The statement of the problem is simplified.
The problem is repeated once more. The investigator draws the patient's attention to what is given and what should be learned in the problem. "The son is five-years old and it is given how old his father will be in 15 years. No, . . . it is said that he will be three times older than the son. Generally. . . well. . . we should learn how old the father is now."	After an additional encouragement the patient correctly repeats the statement, overcoming echolalia.

Draw up the solution plan. "How old is the father now... I'll learn." *Can you learn this in one go?* "No, I can't...

There are several... what do you call it... questions here." *What does the statement say?*

" It is known how old the child is." *How old is he then?* "Five-years old..." *What else is known?* "In 15 years . . . will be . . . three times older than the child." *Solve the problem.* "It does not hang together..."

The patient can't solve the problem mentally. He is asked to do it in writing

The patient writes: "5 + 15 = 20; 15 + 20 = ... (crosses the second operation out) 15 + 15 = 30" *What do you learn by this operation?* "How old the father will be in two 15-year periods."

Have you been solving it correctly? "No, incorrectly." *Then why are you doing* it? "I don't know myself... I cannot think... My head keeps silent."

Write down the correct solution. He writes: "20 × 3 = 60." *Have you solved the problem?* "No." *What should you learn next?* "How old the father is now." *How will you learn this?* "2 x 15 = 30." *Why did you multiply two by 15?* "2×15 . . . I'll cross it out... okay... that's what it should be like.(writes) 60−(2 × 15)=30" *Why did you deduce two by 15?* "Because in 15 years... well, really, there is something wrong here." *When will be the father 60 years old?* "Yes, that's right, in 15 years . . . So, 60−15=45. The father is 45-years old now."

He cannot develop the solution plan unaided. After an additional encouragement the patient notes that the algorithm consists of several questions.

Active solution is inhibited.

The patient begins to perform the operation correctly but is immediately affected by perseveration errors.

He critically evaluates the incorrect persevering "solution" elements.

The patient continues to solve the problem correctly but slides to the inert operation with number 15 which is multiplied by two (twice as old).

The problem is solved correctly after additional support of the solution algorithm.

This dialogue illustrates to what extent the fulfillment of the plan, charted by the patient, is hindered by the pathological inertness of established stereotypes. It will be shown below that only the splitting of the problem into a cycle of successive questions, enables him to eliminate the inadequate operations resulting from the inert stereotypes, and to solve the problem correctly.

(6) We have highlighted the difficulty faced by the patient in moving to a problem with a more complicated solution algorithm that requires the identification of an intermediate unknown value with the subsequent inverted operation order.

Now we are going to review the data obtained after the further complication of intellectual operations.

Such a complication may result either from going over to the *standard* problems that require the introduction of a special intermediate operation or from providing the *conflicting conditions* when the solution algorithm challenges a well-known but inadequate operation.

The patient is given the following problem: x + y = a, provided y = +/−n or y = n(x)

This standard problem can be solved exclusively with the help of a special additional operation: either the preliminary discarding of the remainder (+/− n) followed by the halving of the remaining value and the addition of the discarded number, or the preliminary calculation of the number of "parts" with the subsequent computation of the distribution elements. In both cases the individual should inhibit the direct impulse to halve the sum (as there are two components in the problem) and choose a more complicated solution algorithm.

This was especially difficult for our patient who had had these abilities in his previous experience but had lost them as a result of the pathological inertness of the impulsive stereotypes.

Here follows the description of specific data.

The patient receives the problem: *There were 18 books on two shelves but they were unequally distributed: there were two books more on one of the shelves. How many books were there on each shelf?*

The patient repeats the statement correctly, overcoming typical perseveration errors in the course of repetition.

Tell me your solution plan. "The solution plan . . . the solution plan. . . Well, I'll first add. . . in general, two. . ." *To what will you add?* "I'd add to nine." *And where did you get nine from?* "I got nine from the statement." *Repeat the statement.*" No, nine was not mentioned in the statement. . . I divided 18. . . and added two. . . that will make seven and eleven."

The patient mentally performs the impulsive operation of halving the sum and tries to add the remainder to the obtained result. He is unaware of the performed operation.

The statement is read again. The patient is requested to repeat it. He repeats the statement correctly

The question is replaced by the repetition of the statement data.

What should we learn? "How many there were on two shelves. . ." *Think, what should you learn.* "How many books were there on two shelves." *And what is known in the problem?* "Oh! It is already known how many books there were on two shelves. . . We should learn separately seven and eleven."

The same.

The question is put correctly after the comparison with the statement.

How do you learn this? "I first added seven. . . that is, two to nine. . . and then subtracted seven from 18 and got eleven." *Did you solve the problem*

The inert solution algorithm changes the original statement.

correctly? "Yes, I did." *What is the difference between 11 and seven?* "Four." *And what should be the difference, according to the statement?* "Probably, also four."

The obtained result (7 and 11) is turned into the question.

The patient is prompted the conventional technique. *You should first discard the extra books.* "Then I should first discard the extra books . . . So, 18 − 2 = 16... then halve... that will make eight... 8 + 8 = 16." *What did you do?* "No, I got equal parts... then added 8 + 2 = 10... So, eight and ten equals exactly eighteen."

Having followed the correct solution procedure the patient performs the stereotyped unnecessary operation of addition.

Having used the suggested technique, the patient successfully transfers it to other similar problems although he repeatedly tends first to perform inertly the impulsive operation of halving the sum total but then, after additional stimulation, overcomes this inert stereotype.

It should be noted that when the patient is given a problem with a similar content but a different solution algorithm, he also tends first to use the ready-made stereotyped solution. But then, however, this stereotype is overcome by the patient and the problem is solved correctly.

Here follows another illustration.

After the patient had solved the above problem he was at once given a new one: *There were 21 books on two bookshelves but they were unequally distributed: on one of the shelves there were twice as many books as on the other. How many books were there on each shelf.*

The patient repeats the statement correctly. *Do you know how to solve this problem?* "Not yet." *Try to recall the statement and solve the problem.* "I'll first subtract twice as many... In general, it seven and fourteen." *How did you learn this?* "I... divided into three... that makes seven. *And what then?* "Then we'd add seven twice . . . that will make 17 and then we'll get 21." *Why did you triparpartite?* "Tripartite... but because it is twice as many... So, we should tripartite. Seven is one part . . . " *Does this problem resemble the previous one?* "No, it does There were two more books there and there are twice as many books here."

The patient begins with the wording that ruins the previous and the adequate operations (addition and multiplication, respectively). But then the operation is performed correctly despite some perseveration errors in numbers.

Thus, the patient is able to switch to the algorithm of the problem, which is altogether different from the previous one and he can overcome the inertness of the once-established stereotype that prevents him from resolving the subsequent problem.

(7) Let us, finally, analyze a problem whose semantic solution algorithm is fraught with the conflict between the correct solution methods and the

well-secured stereotype resulting from the direct perception of some statement elements.

In terms of its formal structure, this problem does not differ much from the relatively difficult composite problems we have discussed so far. But it is especially difficult for the patient, due to its inherent psychological conflict.

The following problem serves as an illustration of this case.

The shadow from a pencil which is 16 cm long is 48 cm longer than the pencil itself. How many times is the shadow longer than the pencil?

This problem is distinguished by the confrontation of two operations: the addition of the specific values in the first part of the problem and the division in its second part. Another specific feature of this problem is that the second constituent (the length of the shadow) is made up of a specific material value (the length of the pencil) to which an abstract value (the difference) is added. The final difficulty inherent in the problem is that the impulsive perception of the statement, leads to the tendency to simplify the statement and simply divide the second value (48 cm) into the first one (16 cm).

We observed that the patient faced serious difficulty in solving this problem as a result of the above factors. But in contrast to patients with massive lesions of the frontal lobes, to be discussed below, he was able to overcome these difficulties and solve the problem correctly with the help of external supports.

Let us review how he did it.

Repeat the problem. "The pencil is 16 cm long... it casts a shadow... which is 48 cm long..." *What should we learn?* "How much longer is the shadow than the pencil?

The problem is repeated again. "A pencil which is 16 cm long casts a shadow It is 48 cm longer than the pencil. We should learn how much shorter is the pencil than its shadow." *Did you repeat the problem correctly?* "No, I didn't."

The patient simplifies the statement turning a relative value (48 cm longer) into an absolute one (48 cm long). The first simplification is eliminated but the ratio "48 cm longer" is transferred to the question of the problem. The patient is aware of the incorrectness of the inert wording.

The problem is read in parts. The patient repeats it correctly.

Draw up the solution plan. (He begins to solve the problem at once.) "48 ÷ 16... no, 48 − 16." *What do you learn by it?* "What do I learn? How much longer is the shadow than the pen..." *What then?* "That's all!" When the problem was repeated once more he again phrased the question of the problem as "how much shorter is the pencil than its shadow?"

The complex algorithm is replaced by fragmentary operations. The simplified question is inertly repeated again.

While repeating the statement and solving the problem, the patient apparently faced two difficulties. The first one resulted from the simplified understanding of the statement and replacement of a relative value ("48 cm longer") by an absolute one ("48 cm"). This difficulty was easily eliminated by the patient. The other difficulty was caused by his inability to move from the operation of adding the lengths ("48 cm longer") to that of dividing into parts ("how many times more"). It turned out that it was very difficult for the patient to overcome this difficulty as he repeatedly returned to the changed wording of the question ("How many cm longer is the shadow?") The consistent return to this phrasing testifies to the pathological inertness of established stereotypes which is generally typical of this group of patients.

The patient's main defect described above, that is, his inactivity and pathological inertness, is manifested in this case both as a tendency to substitute fragmentary operations for the complex searching activity and as a difficulty faced by the patient in going from one system of operations (the addition of the lengths) to another one (division into parts) required in the problem.

The patient can, however, overcome these difficulties and successfully solve the problem relying on the written statement, additional encouragement by the investigator, and the successive splitting of the operations.

But it is important to note, that the elimination of the difficulties encountered by the patient in the case of this most complicated problem is unstable, and he is soon affected by the above factors which prevent him from solving the problem correctly.

A week after the first test the same problem was given to the patient once again.

The patient repeats the statement correctly but alters the question which he recites as "How much longer is the shadow than the pencil?" After the problem is read again the patient begins to formulate this question as "How much times longer is the shadow than the pencil?" Only after another reading is the question properly formulated.

The pathological inertness of the formula "48 cm longer" leads to the complete alteration of the final question of the problem followed by its contaminated reproduction.

Tell me how you are planning to solve the problem. Can you solve it in one go? "No, I can't... we should first learn... how many is the length of the shadow longer than the pencil... 48 − 16" *And what then?* "That will be the remainder of the shadow . . . " *And what then?* "And then . . . we shall learn the length of the shadow... 48 − 16... that's all." *Did you solve the problem?* "It seems to me I did" *What should you learn in the problem?* "How much longer is the pencil... I forgot it."

We witness here the direct effect of subtraction as a result of the association between the verbal construction ("how much longer, shorter, etc.") and the operation of subtraction which is well-secured in the patient's previous experience.

The final question gets lost.

The investigator reminds the patient of the question. "Oh . . . to halve . . . no, not to halve but divide into 16!" *What are you going to divide into 16?*

"The remainder of the shadow... The shadow cast by the pencil should be multiplied . . . that is, divided into three and plus two..."

The effect of the inert stereotype of the previous problem in parts.

The further splitting of the problem with the use of auxiliary schemes turns out to be insufficient to overcome the impulsive attempts at subtraction (48 − 16) or division (48 ÷ 16). Therefore, the problem remains unsolved.

The conflicting nature of the problem (and the need to eliminate the direct operations of division or subtraction and to adopt a more complicated solution algorithm) is inaccessible to the patient who suffers from pathological inertness of established stereotypes. It is precisely these circumstances that enable us to graphically see the limits of his intellectual potential.

Let us consider another important issue.

As noted above, the pathological inertness of established stereotypes, characteristic of this patient, revealed itself differently in different intellectual operations. It was inconspicuous in visual operations, for instance, in reproducing the presented movements, naming objects, repeating separate words or performing individual logical operations. It grew, however, increasingly more marked when the patient began to perform serial processes or had to perform operations from memory. The pathological inertness of established stereotypes grew even more apparent if he had to actively analyze the given statement, distinguish the essential relationships, look for an adequate operation program and determine respective operations. The facts, illustrating this statement, were cited above. Hence, there arises the main question: *do all the manifestations of the pathological inertness of established stereotypes persist if there is no need to look for the proper program and the patient could simply use a ready-made (although sufficiently complicated) algorithm?* Psychologically, we may also formulate this question as follows: do the above pathological features of higher cortical processes reveal themselves only in the patient's active intellectual activity or are they equally apparent in the performance of ready-made intellectual operations?

To answer this question we conducted a series of control tests with the patient.

(8) After the patient had completely failed to solve a certain problem or found it extremely difficult to solve it he was given *a ready-made scheme of the solution algorithm and was asked simply to perform the respective operations.*

The patient, as a rule, did not encounter any serious difficulty in performing these operations and very soon gave the correct answer. Furthermore, he overcame the arising perseveration errors comparatively easily. Here follow the respective illustrations.

The solution of a problem that requires the search for the algorithm.

The execution of the ready-made algorithm.

A son is five-years old. In 15 years his father will be three times older than the son. How old is the father now?

If we increase five by 15 units and increase the resultant sum three times we shall derive a number which is 15 units more than the desired unknown. Find the desired number.

"Five . . . so, I'll multiply by 15." *Think better.* "No, I'll add . . . And then I'll multiply by three . . . That will make 60 . . . and then I'll subtract again . . . It will make 45."

There were 18 books on two shelves; on one of them there were two more books than on the other shelf.

You should split the number 18 in such a way that the second value should be two more units than the first one.

"Eighteen should be halved . . . no, I'm saying something wrong . . . I'll . . ." *Write what you want to do.* "18 − 2 = 16 . . . Eight and ten." *What did you do?* "I halved 16 and then added two."

A pencil is 16 cm long. Its shadow is 48 cm longer. How many times is the shadow longer than the pencil?

If we increase number 16 by 48 we'll get a number which is several times more than 16. How many times is the desired number more than 16? The patient immediately writes the solution of the problem. "16 + 48 = 64; 64 ÷ 4 = 4, that is, 64 ÷ 16 = 4." The answer is given without being distracted or replacing the proper solution procedure by impulsive operations.

The provided facts have graphically proven that the main difficulty encountered by the patient in the course of intellectual activity is caused not by the *performance of the ready-made algorithms but by the need for an active search for the programs, that necessitates the preliminary analysis of the statement and the distinguishing of the required operation algorithm from the numerous alternatives.*

The main difficulty faced by people with injured frontal lobes must lie in the impairment of this type of activity, that is, under such conditions all the above impairments grow especially conspicuous.

Conclusions

We have described the form of intellectual disorder that drastically differs both from the intellectual impairments in the lesion of posterior (gnostic) cortical areas and the lesion of the basal-frontal sections of the brain. They also differ a lot from the intellectual disturbances caused by the prefrontal lesions to be described below.

The patient, whose case was used here as an illustration, enjoyed intact intentions, underlying intellectual processes, and preserved programs, into which these intentions are translated. He neither faced difficulty in performing the required operations observed in the case of people with the parietal-occipital lesions, nor suffered from the impulsiveness typical of patients with injured basal sections of the frontal area.

The main defect characteristic of this patient (similar to some other patients with the posterior-frontal lesions) lies in his *inactivity* and the *pathological inertness of established stereotypes* that hinder the normal actualization of his intellectual processes.

These main defects result in the above clinical picture.

The patient repeats all the problems given to him well. He solves problems with a sufficiently simple solution algorithm adequately, as well as encountering no difficulty in solving problems depending on a system of relationships, well secured in his previous experience. But he begins to encounter evident difficulty whenever the statement requires preliminary orientation:

- in the elements which are explicitly formulated in it,
- in the language of additional questions,
- in the performance of intermediate operations which are simply implied in the statement.

The patient's inactivity and the pathological inertness of his nervous processes lead in these cases to the simplification of the solution scheme: a simplified or stereotyped solution substitutes for the new operations that can result only from an active orientation of the problem statement. The patient usually replaces an active search for the adequate solution scheme by the persevering repetition of the applied solution methods. He is unable to correct them although he feels their inadequacy.

Therefore, it is clear that the splitting of a problem into a series of consecutive operations and a system of visual supports help the patient to eliminate his impairments and perform the required intellectual operations.

These are the main specific features of intellectual disorders caused by the syndrome of the lesion of the posterior-frontal sections of the left hemisphere. They distinguish this syndrome from the severe structural impairments of intellectual processes caused by the massive lesions of the prefrontal cortical sections.

---

[1] These cases will be considered in Chapters 5 and 6.

Chapter 5

DISORDERS OF PROBLEM SOLVING IN THE DEVELOPMENT OF THE MASSIVE FRONTAL SYNDROME

Our previous analysis enabled us to distinguish two factors leading to intellectual disorders in different variants of the frontal syndrome. The first factor was the pathological weakening of higher inhibition processes, and the replacement of preliminary analytical-synthetic activity with impulsive attempts at problem solving. This factor could be observed in different injuries to the frontal lobes, but was especially conspicuous in the lesion of their basal areas. The ensuing symptomatic picture of an intellectual disorder emerged, against a background of typical emotional-volitional impairments and the potentially intact formal intellectual operations.

The second possible factor accounting for intellectual disorder, inactivity and the pathological inertness of established stereotypes, could accompany lesions of the frontal lobes localized to different areas. It was, however, most conspicuous in the lesion of the posterior-frontal areas of the left hemisphere. It is important to note that the pathological inertness of established stereotypes was especially manifest in speech and that it inhibited potentially intact intellectual operations. But the use of special methods designed to give the patient external supports and eliminate the need for active searching successfully balanced the defective intellectual processes.

We have already described the two variants of intellectual disorders accompanying injuries to the different sections of the frontal lobes. Our present objective is to trace *the precise impact of a gradually evolving injury to the different systems of the frontal areas on the syndrome's development,* as well as to examine the emerging intellectual disturbances in the case of *one and the same patient,* observed at successive stages of the disease.

Such an opportunity was provided by the case of patient U. His slow growing intercerebral tumor in the right frontal lobe required its resection, after which the tumor spread to the left frontal lobe and partially destroyed it.

Let us analyze this extremely important series of experiments.

Patient U: General Information

Patient U, 30-years-old, a foreman with ten years of schooling, was first admitted to the Institute of Neurosurgery in 1960 (case history No. 29,302) complaining of "attacks" which began with general weakness and the sensation of "heat" in the epigastric area which spread upwards and reached his head. He had been suffering from such attacks for two or three years. In January 1960 he began to suffer from headaches which were accompanied by dizziness and vomiting; his vision decreased and the examination noted papilledema in both eyes. The EEG showed right hemisphere slow wave activity. The aggravation of his condition necessitated ventriculography which revealed hydrocephalus of the left lateral ventricle and the underfilling of the right one. The liquid was hydrocephalic. The patient's severe condition necessitated the decompression of the right frontal-temporal area that resulted in a general improvement. Some time later, however, we noted the encephalocele and the left-sided hemiparesis, more manifest in his arm and hand. His papilledema grew more conspicuous, and the protein contents in the liquid increased. Intracranial pressure reached the level of 270 mm. As it seemed a massive process was under way in the right frontal-temporal area, the patient underwent another operation: the right lateral ventricle was opened but no tumor was found. After the patient's condition had sufficiently improved he was transferred to the Institute of Neurosurgery.

The clinical picture observed there was very unclear. The patient complained of nothing special. In the fundus of his eyes we noted colorless optic papillae although his vision was normally acute and the fields of vision preserved. We observed certain hyposmia to the right but did not note any sensitivity or motor impairments. We registered hyperactive tendon reflexes to the right. The patient's EEG showed the focus of pathological activity in the right frontal-anterior-temporal area (the area that was operated on).

The neuropsychological assessment conducted at that time by E.D. Kholmskaya testified to the *patient's intellectual intactness*. There were no impairments either in his gnosis or in praxis and speech. Against this background we noted, however, apparent disturbances in reciprocal coordination, mild perseveration errors on tests of serially organized movements and some retardation in solving intellectual problems, which evidently strained him.

All this data led to the following provisional diagnosis: a parasagittal tumor located deep in the anterior sections of the right hemisphere; probably, a tumor of the transparent septum. As the patient's condition at that time was rather good, we decided against another operation.

Two months after his discharge from hospital, his condition drastically deteriorated: he began to experience a feeling of cold and growing weakness in his left side. General weakness and somnolence became more apparent. He once again began to suffer from epileptic attacks, accompanied by the feeling of "not enough air" and vegetative symptoms, followed by loss of consciousness and general convulsions.

He was readmitted to the Institute of Neurosurgery (case history No. 30,335). The patient was still completely oriented, adequate and did not exhibit any apparent mental disorders. We could observe only some manifestations of growing weakness in his left arm with a tremor of both hands; the symptoms of Oppenheim and Gordon signs to the left; a hint at the grasp reflex to the left; and a mild palmomental reflex. The EEG exhibited the increasing general cerebral symptoms. Against this background, we noted a focus of pathological activity in the right frontal area which grew even more severe than before. The arteriogram showed a significant antedisplacement of the anterior cerebral artery. It was flat and extended in its anterior extremity, which testified to a tumor in the basal-medial areas of the right frontal lobe located very close to the lateral ventricle and the transparent septum.

This information confirmed our previous suggestion, and on October 23, 1960 the patient was operated on by Dr. Abaimova, who coagulated the cortex on the boundary of the middle and superior frontal convolutions to the right, and found a tumor located at a depth of 4 cm in the basal-medial areas of the right frontal lobe. The tumor grew into the anterior horn of the right lateral ventricle. The tumor with very few vessels was dense and elastic (histologically — an astrocytoma). It was removed in parts with the help of an electric loop. A small part of the tumor in the medial sections of the frontal-basal area was not removed.

After the post-operation period, which featured no complications, the patient was discharged from hospital. His condition at the time of discharge was quite good. We noted some manifestations of left-side hemiparesis but no apparent intellectual impairments. During the last days of his stay at hospital we observed only *slight emotional changes* in the form of mild euphoria, disinhibition and talkativeness against a background of general retardation and languor. All these symptoms revealed themselves against a background of sufficiently intact gnosis, praxis, speech and formal intellectual processes.

After discharge from hospital the patient resumed his work but soon left his job. The patient gradually began to show manifest mental changes: he grew languid, inactive, retarded and sleepy. He stayed most of the day in bed saying that he "did not feel like doing anything." He read nothing pleading a "shortage of interesting books." He was very talkative and used many stereotyped constructions in his speech. In conversation he generally displayed a lack of initiative but passively answered questions. According to his wife, by mid-1962 he grew so inactive that he needed additional encouragement to perform any task: he could freeze, having begun a task, and discontinue it. While standing up from bed he could freeze in any intermediate position and needed additional instructions to move from one position to another.

In March and April the epileptic attacks with tonic convulsions in the left extremities and the loss of consciousness resumed. As a result of the apparently aggravating symptoms in June 1962, the patient was for the third time readmitted to the Institute of Neurosurgery (case history No. 33,847). The evident aggravation of symptoms testified to the continued growth of the tumor:

- engorged papillae on a background of a concentric narrowing of the vision field margins which was more conspicuous to the right;
- anosmia to the right and hyposmia to the left;
- paresis of upward gaze;
- severe 7th cranial nerve paralysis on the left;
- muscular hypertension in the left arm evident against early manifestations of paresis in the right arm and hand;
- bilateral pathological reflexes especially marked to the left with definite plastic hypertension;
- palmomental reflex to the left and an apparent lip reflex.

There were, however, no disturbances of sensation.

The patient's neuropsychological assessment noted general inhibition, languor and lack of dynamism. All movements were extremely retarded and inhibited. His face was expressionless and speech monotonous. Alongside these manifestations we noted euphoria and distractability. At this time he had become much less critical of his condition.

Against a background of general adynamia and an evident apathetic-abulic syndrome we noted impairments in reciprocal coordination and unnecessary beats in the rhythm tests which the patient could not inhibit. His speech was monotonous and retarded; the intellectual operations were still formally intact but they were inactively actualized with marked dynamic difficulties (this point will be specifically elaborated below).

The electroencephalogram showed an apparent focus of pathological activity in the right frontal-temporal area against a background of severe general cerebral changes.

The arteriogram displayed a severe antedisplacement and sinistroposition of the right anterior cerebral artery and the enlarged system of vessels in the basal-frontal area, which testified to a large tumor in the basal sections of the right frontal area.

All the symptoms testified to the continued growth of the tumor, which exerted pressure on the subcortical ganglions and affected secondarily the olfactory tracts against a background of high intracranial pressure with the secondary lamina tecti signs.

The rapidly growing symptoms necessitated a *second operation.*[1] The patient was operated on by Dr. Abaimova on June 27, 1962.

After the dura mater encephali had been cut along the old paunch, the extremely thinned cortex of the right frontal lobe was coagulated and cut, revealing a thick, very large tumor with evident boundaries. *The tumor occupied the entire right lobe from the pole to its posterior sections and from the convex cortical areas to the base of the anterior cranial fossa.* In its outer sections it penetrated the anterior-medial sections of the temporal lobe, reaching the falx and spreading underneath to the left. The removal of the tumor left an enormous cavity (12 x 12 x 10). Its walls were formed from the front by the medial falx, from the base — by the dura mater encephali of the anterior cranial fossa, from the outside — by the greatly changed and thinned brain tissue and from behind and the top — by the brain tissue. The medial-posterior areas of the tumor, which spread under the falx, were not removed but its outstanding section was coagulated. The frontal cerebral artery located in the medial area of the tumor was clipped. Histologically, the tumor, with very few vessels, was an astrocytoma.

The post-operative period was characterized by a severe aggravation of symptoms. The patient began to suffer from hallucinations and mental confusion. He was untidy in bed, woke up from the sensation of cold, stood up and tried to find a vacant bed in order to occupy it. When his wife visited him the patient told her that he had given her a car and asked how she drove it. Later he forgot all about these conditions. The hallucinations and confusion gradually disappeared but severe behavioral changes persisted. The patient was disinhibited, sexual, took liberties with the nurses, made cynical remarks. His behavior could hardly be controlled. Six weeks after the operation these symptoms gradually began to disappear. The patient grew calmer but complacency and an uncritical attitude persisted. The symptoms of higher cortical impairments also grew much less evident: his movements were quicker, the disturbances in reciprocal coordination gradually disappeared, tests on the serial movements were performed with very few impairments. He could already reproduce the rhythms, unnecessary impulsive movements and perseverations were practically inconspicuous. Formal intellectual operations remained intact although they still featured some manifestations of impulsiveness and perseverations.

The patient was discharged from hospital on August 7, 1962. His condition at discharge from hospital was satisfactory. But in November 1962 his condition began to measurably deteriorate:

- there developed marked changes in his behavior;
- he grew aggressive: he began to bring his fists into play when approached by other people and could not inhibit this urge;
- he grew more languid, less spontaneous;
- he spent most of his time in bed;
- the patient grew untidy;
- epileptic attacks resumed; they were accompanied by cramps and the turn of the head to the left.

The patient was readmitted to the Institute of Neurosurgery on April 15, 1963 on the suspicion of continued tumor growth (case history No. 35,673).

His assessment noted symptoms testifying to the continual tumor growth (such as bilateral hyperactive pathological reflexes, metabolic disorders [obesity]) against a background of the above neurological manifestations, which did not measurably change as compared with the clinical picture before the second operation.

The patient's neuropsychological assessment noted drastic changes in his general behavior. He suffered from the complete loss of spontaneity, dynamism and initiative. He did not request anything and gave only "yes/no" answers to questions addressed to him. His contribution to a conversation was limited to formal remarks. The patient was very uncritical of his condition: he said that he had not changed at all during his illness and "remained precisely what he was before."

Therefore, we suggested that the tumor continued to grow, this time spreading to the third ventricle and to the left hemisphere, and compressed the left frontal lobe.

On May 8, 1963 the patient was operated on for the third time by Dr. Abaimova who opened the dura mater encephali in the frontal area of the right hemisphere and found a large cavity, described above. It is noteworthy that there was a tumor covered by a dense grey film in the posterior medial sections. After the film had been cut along the lower edge of the falx the surgeon found both on the outside and below near the bottom of the anterior cranial fossa a protrusion of the tumor of a pink-meaty color. The tumor was separated from the base of the anterior fossa and partly removed, revealing an apparent spreading of the tumor under the falx to the left hemisphere. The section of the tumor spreading to the left hemisphere was not removed. Histologically, the tumor was still an astrocytoma.

The post-operative period went smoothly. The patient still exhibited a severe lack of spontaneity and initiative, incontinence, pleading that he had no time to ask for a bed-pan. He remained insufficiently critical, exhibiting manifestations of euphoria. The patient was very talkative and his speech was still full of stereotypes.

The patient's neuropsychological assessment conducted by Prof. Luria two weeks after the operation noted the following picture:

The patient is well aware of place and time but severely aspontaneous. He does not formulate any complaints or requests himself, and cannot coherently retell his case history. He is, however, able to answer questions and tell his recent life-story accurately. The patient was highly distractable; he may be diverted by any irritant: he discontinues his work and begins to perform it worse. He remembers well experiments in which he participated, retells the summary of stories read to him a month before and recites their maxims.

He has neither any apparent disturbances in position praxis nor any spatial praxis disorders, but suffers from obvious impairments in dynamic praxis. He encounters difficulty in performing tests on reciprocal coordination, easily loses his motor set, and begins performing identical movements by both arms. While imitating rhythms presented by ear, the patient easily loses his motor set and begins producing rhythms. When reproducing rhythms on a verbal instruction, he usually substitutes an inert reproduction of an established stereotype for the desired rhythm.

His speech is perfectly intact; he easily repeats words, names objects, and carries out simple directions. But when the length of a word series increases to five words he begins, however, to experience difficulty going from one word series to another and replaces a new series by inertly reproduced stereotypes.

The patient's intellectual operations reveal similar dynamic impairments. He repeats a story he has read easily and is able to grasp its idea. The patient can count well but he exhibits an apparent inactivity with a marked tendency to use inert established stereotypes. These intellectual impairments will be specifically dealt with below.

Since the part of the tumor in the left hemisphere was not removed, it was decided to perform another operation. This time he underwent surgery in his left frontal lobe.

On June 27, 1963 the patient was operated on for the *fourth* time by Dr. Abaimova who cut the dura mater encephali of the left frontal area. The convolutions of the brain were flat and did not pulsate. Having punctured the left frontal lobe towards the lateral ventricle, the surgeon cut the cortex of the left frontal area in parallel to the median line and found a *well-encapsulated tumor* at a depth of 4 cm. The tumor was removed by a pointed curette. Its removal revealed a cavity in the lateral ventricle. The tumor extended down to the third ventricle. The bulk of the tumor was located *in the medial sections of the left frontal lobe and spread medially.* It could be easily separated from the lateral sections of the frontal lobe and spread basally inward, penetrating the cortex at the basis of the left frontal lobe. Its removal revealed the anterior cranial fossa and the left optic nerve. The part of the tumor, still located medially, was not removed. Histologically, the tumor was an oligoastrocytoma. Its removal left a cavity 10 x 10 x 12 cm in size.

The post-operative period went smoothly. In the first days after the operation the patient was unaware of place and time but quickly recovered spatial orientation. This condition did not cause any marked neurological symptoms, but we noted stable psychic changes which persisted for the rest of the follow-up period. His behavior drastically changed. He became rude, laughed loudly and did not realize that he disturbed his neighbors in the ward. But throughout the first weeks after the operation, he remained sufficiently oriented in space and time.

The manifestations of aspontaneity and adynamia, described before the fourth operation, persisted. The patient was still languid but he easily entered into conversation. He often answered questions by formal remarks and stereotypes. For instance, he said that "it was March" and did not notice that it contradicted his earlier remark that he was admitted to hospital on April 15. He easily began to work out the received assignments but got immediately exhausted and substituted inert stereotypes for the correctly performed tasks. He still did not suffer from any apparent disorders in spatial praxis or position praxis, but in tests on dynamic praxis he very quickly lost the given program of movements and replaced it either with inert stereotypes or disordered motor actions. We found that the patient encountered difficulty in memorizing complex motor responses and that the learned complex motor responses were easily forgotten. His speech was preserved, but we noted much more serious difficulties in remembering the given programs, and much more manifest pathological inertness in performing intellectual operations following his most recent surgery.

The described manifestations persisted throughout the subsequent two-months' follow-up period. The patient remained aspontaneous, complacent and uncritical. He did not show the proper emotional attitude to his defects, and was euphoric and light-hearted. The patient agreed to work with the therapist but did not correct his mistakes. His most severe intellectual disorders did not generate any noticeable emotional reactions.

Toward the end of the second post-operative month, the manifestations of general disinhibition began to fade, but severe intellectual impairments persisted, against a background of general impairments in purposeful behavior and an uncritical attitude to his condition.

The review of this case is extremely interesting: we could clearly observe *three different syndromes* with successively aggravating changes in the *three distinguishable periods* in the case history.

In the *early period* the tumor was restricted to the transparent septum and the walls of the ventricle and the patient's higher cortical functions were perfectly intact. His symptoms were limited to paroxysmal vegetative impairments, described above, and some isolated impairments of reciprocal coordination, that testified to the influence of a pathological process on the motor relations between the hemispheres.

The *first period*, (after the removal of a large tumor from the right frontal lobe during the first radical operation) marked the development of the apparent syndrome of the severe dysfunctions in the *right frontal area.* This section of the brain was first drastically impaired by the tumor and then almost the whole of it was resected. A similar condition also persisted in the subsequent *second period.* It endured partially before and after the second radical operation. This period featured severe affective impairments and the patient's marked inactivity, but he nevertheless retained an ability to control his actions (including intellectual operations) through a well-retained program.

Finally, in the *third period* directly before and after the *third radical operation* the clinical picture altered once again. During this period the tumor had grown into the *left frontal lobe* and was removed with the large part of its matter. This period featured the aggravation of severe disorders of the patient's personality, which took the form of an apparent apathetic-akinetic-abulic syndrome, and very conspicuous intellectual disturbances that revealed themselves by his inability to retain and carry out any complex intellectual programs. The latter syndrome persisted after the general hypertension symptoms had disappeared due to the surgery. It also persisted when the severe impairments in the regulation of higher mental processes began to reveal themselves, even though his physical condition appeared normal.

Thereby, we can describe the *dynamics of the evolving frontal syndrome,* using one case as an illustration, and specifically distinguish the syndrome of the early dysfunctions in the frontal area, which later developed into a severe syndrome of *the bilateral lesion of the frontal lobes.* We shall try to do it analyzing in detail the intellectual changes observed in the three latest periods.

Disorders in Problem Solving in the First and Second Stages of Patient U's Disease

As noted above, in the early stage of his disease (1959–1960) patient U had no apparent higher cortical dysfunctions. His intellectual processes were somewhat retarded but structurally intact. They were goal-oriented. He did not exhibit pathological inertness, pathological impulsiveness or fragmentary intellectual operations. Therefore, this stage is of little interest to us and we are not going to review it in detail.

We are much more concerned about the first and the second stages of his disease (1962–early 1963).

By the first stage the patient suffered from apparent aspontaneity: each element of a complexly structured activity required special additional stimulation. His motor processes began to graphically show the pathological inertness of established stereotypes (for example, the stereotype of the rhythmic beating at one tempo, repeated even after the patient had got another instruction, for instance to beat in two strokes). He did not realize that his responses were incorrect. The patient's mnemic processes also revealed some symptoms of pathological inertness. He could recite a series of four and even five words well, but began to repeat the previous word sequencing if it had been changed. The patient retained, however, an ability to repeat the most common word series in reverse order, could reproduce and remember the content of an intricate story and understood its implied idea. His calculation skills were preserved and he could do simple mental arithmetic without error. But his work on these mathematical problems clearly revealed his inactivity. Furthermore, he needed additional encouragement to perform operations which required mental arithmetic (for instance, $81 - 24$); otherwise he was distracted by any outside stimuli and froze. He still could perform the operations of serial arithmetic (for example, count backwards in sevens) although they were retarded and lacked fluidity. Nevertheless, these operations were only seldom accompanied by perseverations which the patient could eliminate.

It was against this background that we began a careful study of his intellectual activity.

To do this, we once again began the analysis of his problem solving process. Like before, we began with mathematical problems with a relatively simple solution algorithm and progressed to more difficult problems that required a series of consecutive operations and sometimes specifically the elimination of established stereotypes.

(1) Patient U was good at solving *simple problems of the* $a + b = x$ *type or elementary composite problems of the type:* $a + (a + b) = x$. Therefore, we will not discuss them in detail.

He could solve slightly more complicated problems which required division followed by multiplication easily enough. Hence, we shall limit ourselves to one illustration, as it is not necessary to describe this level of intellectual processes in greater detail.

The patient is given the problem: *A housewife used 20 liters of kerosene in five days. How many liters did she use in a week?*

The patient repeats the problem correctly and begins to solve it. "In five days... 20 liters... seven days in the week . . . "(laughing, a pause) *What should you learn?*" How much will she use in two more days . . . Probably, 28 liters . . . "*How did you learn this?*" Four liters a day — in two more days — 8 liters, all in all — 28 liters."

The problem is solved mentally.

The patient is fully aware of the solution process.

This illustration shows that it is very easy for the patient to solve this type of problem and that its solution plan is quite clear to him. It will be shown below that his growing general inactivity, which became increasingly apparent during the subsequent stages of the disease, left the solution algorithm of a problem intact, but necessitated the continuous stimulation of his problem solving process.

(2) We could observe significant impairments in problem solving of more difficult problems when it was necessary to determine two unknown values from two equations and perform many of the operations mentally, into the bargain.

For this, we used problems of the type:

$x + y = a \quad x + 2y = b$; determine the values of x and y.

Our research findings show that the solution of such a problem, that requires the performance of operations from memory, clearly reveals the inertness of established stereotypes. It also revealed a tendency to replace the necessary discourse with the secured "field" operations that graphically expose the inactivity of the patient's intellectual operations.

The patient gets the problem: *One ABC book and one pen cost 37 cents. One ABC book and two pens cost 49 cents. How much do one ABC book and one pen cost separately?*

The patient repeats the problem correctly but does not begin to solve it.

What should you do to solve the problem? "37 and 49... then it's 86 all in all..." *What did you add them for?* "To learn how much they paid all in all." *What should you do next?* "Next . . . I don't know what next... There is nothing else..."

The substitution of a simple summation for the desired operation.

The illustration shows that two components of the statement do not prompt any solution plan to the patient. The question of the problem which cannot be answered directly, is replaced with the question on the sum spent answerable by a simple summation. We see no evidence of transition to the logical reasoning required to solve the problem.

In order to draw the patient's attention to the final question and help him distinguish the main solution elements, the statement is presented in writing in the form of the following equations:

One ABC book + one pen = 37 cents;
One ABC book + two pens = 49 cents.

The question *How much do one ABC book and one pen cost separately?* is repeated.

The patient reads the written equation, draws a line underneath and writes down the result of simple summation: "Two ABC books + three pens = 86 cents," saying: "I've learned how many pens and ABC books, how much everything costs."

It is evident that our attempt to show the difference between the two equations and to prompt the proper logical scheme ended in failure. The discourse was again replaced with a stereotyped operation of summation, which is absolutely inappropriate.

In order to eliminate this stereotype and help the patient formulate the proper solution plan, the investigator gives him the initial element of the respective logical chain.

First 37 cents were spent, then — 49 cents. What is the difference accounted for? "It is accounted for by one pen! Then, 49 − 37 = 1 pen." *What should you learn?* "How much do an ABC book and a pen cost separately." *So?* "So, a pen costs 12 cents. Now an ABC book. . . 49 − 24 = 25 cents. One ABC book costs 25 cents."

The patient easily solves the problem on distinguishing the significant element of the solution algorithm.

Therefore, the patient could not himself initiate the required reasoning process by distinguishing the initial element of the algorithm and was affected by the stereotype (summation) which was irrelevant to the statement of the problem. He could, however, solve the problem easily after the initial element of the algorithm had been distinguished for him.

We should, nevertheless, notice that the successful solution of this problem *did not result in the development of a stable algorithm of respective operations and the patient failed to eliminate the established impulsive stereotype.* Therefore, when he was given a similar problem he could not transfer the previous algorithm and was once again affected by the inert stereotype of direct summation.

The patient is given a similar problem presented in the form of a written equation:

1 kg of salt and 1 kg of apples cost 80 cents.
2 kg of salt and 1 kg of apples cost 90 cents.

How much does one kg of salt and one kg of apples cost separately?

The patient reads the problem, recites it correctly and says: "And how shall I learn it?" *Think a bit.* He shrugs his shoulders: "There is no making head or tail of it."

The statement of the problem does not trigger the discourse.

Write down the problem in one line. The patient writes down the problem in a column and reiterates the stereotyped operation of summation:

1 kg of salt + 1 kg of apples = 80 cents
2 kg of salt + 1 kg of apples = 90 cents
3 kg of salt + 2 kg of apples = $1.70

The patient performs the stereotyped operation of summation although he was shown at the previous problem that it is out of place here.

What did you do this for? (He freezes.) *How will you solve the problem?* I don't know... "*What accounts for the difference in the second case?*" 1 kilogram of salt. "*Solve the problem.*" "3 kilograms of salt... No, I won't be able to work it out till the New Year..."

The distinguishing of the initial element of the solution algorithm does not help. The developed stereotype of summation persists.

What accounts for the difference in money? "1 kilogram of salt... for 10 cents. So, the salt... So, 1 kilogram of salt — 10 cents. 90 − 80 = 10. The apples cost 70 cents."

You solved the problem, didn't you? "And what do you think—is that all? The second problem... Salt 20 cents and apples 70 cents. "*Which answer did you get?*" We should strike a balance: salt — 30 cents, apples 2 kilograms — 1 dollar 40 cents; all in all 1 dollar 70 cents."

What was asked in the problem? "How much do one kilogram of salt and one kilogram of apples separately. Salt — 10 cents, apples — 70 cents." *So you solved the problem, didn't you?* " Did I?! It's due to your help. . .The problems are easy but I can't decide what to begin with on my own."

The problem is solved correctly after the initial element of the algorithm has been distinguished again. The question of the problem is not distinguished as the main one and the further analysis of problem solving is again replaced with the repeated inert stereotype of summation. The correct solution easily crops up after the question has been confirmed.

Hence, the mastered algorithm of problem solving does not eliminate the inert stereotype and the deduction process is again replaced with the inert operation of summation, although it has been illustrated by the previous problem that it is out of place here. It is noteworthy that the correct solution algorithm triggered by prompting persists alongside the inert stereotype of summation. Only another confirmation of the question makes the correct solution algorithm seemingly dominant.

The process of problem solving of the third similar problem reveals the instability of the determined solution plan and the ease with which it is replaced with an inert stereotype.

The patient is given the following problem: *one rod and one fish-hook cost 40 cents; one rod and three fish-hooks cost 60 cents. How much do one rod and one fish-hook cost separately?*

The patient repeats the problem correctly, then writes it down in a column and sums up:

An inert reproduction of the stereotyped summation.

1 rod + 1 fish-hook = 40 cents
1 rod + 3 fish-hooks = 60 cents
2 rods + 3 fish-hooks = 1,000 cents

It is noteworthy that "1,000" in the last number is persevered.

The difference here lies in fish hooks... Then a fish-hook costs 10 cents and a rod — 30 cents."

At the same time the problem is solved correctly.

It is characteristic of this stage of the experiment that *the inert stereotype of summation perseveres alongside the correct solution algorithm* although the redundancy of this operation is self-evident.

A similar situation occurred a day later: the patient repeated the statement of the problem given to him the day before correctly, but persisted with the stereotyped operation of summation, alongside the correct solution of the problem. It is important to note that when the patient was asked why he performed the redundant operation of summation and what he thought about at that time he gave the following answer: " I don't think about anything... It's not to solve the problem... I'm doing it for nothing!"

We offered the patient similar problems many times and the described attitude persisted. He began to solve a problem, the type of which grew quite usual for him, only after the investigator had distinguished the initial element of the solution algorithm by asking the question "What accounted for the difference?"

(3) We obtained similar data for the solution of so-called *standard problems* (of the type: $x + y = a$, provided $y = 2x$).

The patient is given the problem: *There were 18 books on two bookshelves. On one of them there were twice as many books as on the other. How many books were there on each bookshelf?*

The patient repeats it correctly but when he attempts to recite the question he formulates it in the following way: "It seems to me we should learn how many books there were altogether." The patient refuses to solve this problem but after he has been prompted that this is a problem "in parts" he solves it correctly and easily transfers its solution algorithm to another similar problem. But the developed solution technique is inertly reproduced when he is given a problem in difference. The patient is given the following problem: *There were 20 books on two bookshelves but there were two more books on one of them than on the other. How many books were there on each shelf?*

He repeats it correctly and says that this is a problem in parts but immediately adds that he does not know how to solve it. After the second reading the patient notices the difference but he can solve the problem only after he has been given the initial element of the solution algorithm ("discard the difference").

These illustrations prove that the patient suffering from marked inactivity repeats the statement of a given problem well, but in the process of problem solving *easily subordinates his behavior to the established inert stereotype* which replaces the proper solution algorithm. The patient solves the problem correctly when he has been prompted the initial link of the solution algorithm. In this case he *easily finishes the initiated solution process*. But the developed solution stereotype (which contradicts the statement) is so inert that *it perseveres alongside the correct algorithm.*

The patient's thinking at this stage of the disease is characterized by a potentially intact system of correct intellectual operations, which becomes evident after additional stimulation, against a background of apparently pathological inertness of stereotypes irrelevant for this system.

The further evolution of U.'s disease was caused by two factors:

1. the removal of the tumor located within the right frontal area with the resection of virtually the entire right frontal lobe, on one hand;
2. the continued growth of the tumor which spread to the left frontal lobe, on the other.

Clinically, they revealed themselves in the patient's growing general adynamia and aspontaneity accompanied by manifestations of continual loss of criticism and the development of uncontrolled impulses that drastically altered his behavior.

We resumed our observations of the patient's intellectual activity two weeks after the operation. Therefore, we could observe changes in his behavior in the conditions of a special psychological experiment.

This time, with a view to comparing his intellectual processes with our earlier research findings, we did not give the patient any new arithmetic problems. We limited ourselves to the problems that he had already successfully solved.

We came to the conclusion that by then — as a result of his growing inactivity — *he was unable to work out any problems on his own*. We observed fragmented numerical operations which were irrelevant to the system and undetermined by the general solution algorithm of the problem. The patient, however, *retained an ability to solve a problem if it was split into separate questions and all his activity was limited to answers to individual questions.*

These conclusions will be illustrated by the forthcoming extracts from the protocols.

(4) The patient is given problems in parts which require division followed by multiplication. He used to solve them quite well. Now the process of problem solving looks different.

During the third post-operative week the patient gets the problem: *A housewife uses 20 liters of milk in five days. How much milk will she use in a week?*

He repeats the problem easily but cannot do anything to solve it on his own *How will you solve the problem?* "I don't know it myself." *What should you do to work it out?* (Silence) *What should learn?* "How much does she use per day." *And then?* "How much does she use a week!"	The solution algorithm is preserved; the patient is unable to use it unaided. The patient easily finishes the statement of the

Begin to solve the problem. " It's difficult to begin." *A housewife uses...* "20 liters of milk in five days. How much will she use in a week. . ."

problem but does not begin to solve it.

Get down to work. "I've already forgotten how to solve it." *What should you learn?* "How many liters does she use a day." *And then?* "How many in a week." *Solve it.* (A pause, the patient freezes.) *A housewife uses . . .* 20 liters of milk in five days." *Therefore...* "4 liters per day." *Consequently...* "she will use 7 x 4 = 28 in a week."

The problem is solved after additional encouragement.

This illustration testifies to the preserved solution algorithm, but its intactness does not result in the active performance of the required operations. With a view to overcoming this inactivity and revealing the patient's preserved abilities, in the following tests we split the problem into parts and preceded each part with a specific question. The patient begins to work out the problem with a series of answers to successively presented questions.

According to our observations, this method remained accessible to the patient; he answered consecutive questions correctly but was, however, unable to contract this chain of operations and turn it into a contracted independent series of operations.

Here follow respective illustrations.

The patient gets the problem: *A housewife uses six liters of milk in five days. How much will she use in a month?*

The patient easily repeats the problem. When asked to solve it he immediately answers: "30 liters." *How did you learn it?* "5 × 5 = 30" The statement is written down in the form of a table:

5 days—6 liters
30 days—?

The solution is replaced with a direct fragmentary operation with the numbers given in the statement.

The patient immediately says: "This is a problem in parts." *How many five-day periods are there in a month?* "Six" *Hence how much does she use in a month?* "Six times. . . " Therefore? (a pause) *How many five-day periods are there in a month?* "Six" *How much does she use in one five-day period?* "Six liters." *How much does she use in a month?* "36 liters."

The problem is solved through answers to successively presented questions.

The patient is given a similar control problem: *A cow yields seven liters of milk in three days. How much milk does it yield in twelve days?*

Independent solution is blocked.

The patient repeats the problem correctly; when asked to solve it he repeats the statement again but does not get down to work.

Is this problem similar to the previous one? "Yes, it is." *Then how shall we solve it?* (He repeats the statement again but does not begin to solve the problem.) *What should be done?* "I don't know." *Solve the problem.* (a pause)

The patient notes their similarity but this does not lead to the active realization of the solution algorithm.

The patient is asked a series of successive questions. *How much milk does it yield in three days?* "Seven liters." *How many three-day periods are there in 12 days?* "Four" *How much does it yield in 12 days?* "28 liters."

The problem is solved due to individual questions.

The patient was given similar problems during a number of successive sessions. Each time he repeated the problem correctly but did not get down to working it out. Then he wrote down the table used the day before and solved the problem with the help of consecutive questions.

This period of his disease was characterized by his inability to solve a problem unaided, despite the preserved ability to solve it (with the support of a written table and the accompanying separate questions that simply served to encourage the patient).[2]

This period is also characterized by the fact that although the patient could solve a problem correctly if asked a series of consecutive questions, he was often unaware of the method applied and persisted with the point that he had solved the problem " at once." It will be shown below that such impaired analysis of the earlier operations will persist in the subsequent stages of the disease.

Hence, the growing cerebral pathology caused by the continual tumor growth and the resection of the right frontal lobe, led to the increasingly more manifest syndrome of general inactivity. The patient, however, preserved an ability to perform selective intellectual operations determined by the statement, each of which was strictly determined by the respective question from the program offered to the patient.

Changes in the Patient U's Problem Solving in the Third Period of his Disease (after the Partial Resection of the Left Frontal Lobe)

As shown above, the continual growth of the tumor, which spread to the other hemisphere, and the increasingly more apparent symptoms, necessitated the last radical operation. During the operation, a massive tumor was removed from the deep medial sections of the frontal lobe and the measurable section of the left frontal area was resected.

The post-operative period marked continued changes in the symptoms, earlier unobserved outbreaks of uncontrolled behavior, and significant changes in psychological processes.

During the first post-operation weeks the patient was communicative, well aware of space but much less aware of time. For instance, he said that it was May (it was really July) but immediately added: "You know, I stuck on May . . . I underwent an operation in May . . . I have long stopped keeping time . . . " The patient recognized his doctors and nurses but his attitude toward them drastically changed: he grew aggressive, uncontrolled, he sometimes performed meaningless actions — sang, twisted the arms of the attending nurse, etc. From time to time he grew languid, adynamic, inert.

The patient fulfilled the tasks given to him but usually did them very slowly after long pauses.

The tests on his position praxis and spatial praxis conducted at that time noted elements of mirror-like reproduction unobserved before. The tests on consecutive movement series revealed his inability to develop the desired program of movements. The given sequence of motor actions was replaced with haphazard chaotic movements. A respective training course helped him master the required movement sequence which, however, very soon turned into an inert stereotype that prevented the patient from going over to any other series of movements. The patient still could not coordinate the movements of his arms. Their reciprocal coordination was replaced with concomitant movement.

Rhythm tests noted excessive beats and severe stereotypes which could not be eliminated by verbal instructions.

His *visual gnosis* was preserved but the patient encountered serious difficulty in distinguishing individual figures from the background and superimposing the outlines.

His *speech* remained intact; the patient could repeat separate words and name objects well. He was able to enumerate standard rows forward and backward and even name groups of items mentioned in an instruction (five red, three sharp objects, four words beginning with the letter "k", etc.). We did not notice any marked perseverations in his repetition of word series and only more complicated assignments revealed some elements of error. The patient was still able to do simple arithmetic and easily performed such operations as: $53 - 16$; $36 + 4 - 20$; $125 \div 5$; etc. He had neither writing nor reading dysfunctions.

Severe disturbances began to reveal themselves when we conducted experiments in which the patient had to *retell* stories or *develop their plots on his own*. The disturbances manifested themselves, first, in the *minor loss of selectivity of relationships* and, second, *in his absolute inability to develop a plot on his own*.

For example, the patient retold a story by Leo Tolstoy "The Ant and the Pigeon "in which a pigeon saved a drowning ant who repaid her for her service by saving her from a hunter. The patient said: "Once an ant went to the river to drink. . . and was caught by the forest-man(!). A pigeon was flying by, she saw it and says: "So, the old man, that is how you act. . . A day will come when I stand you in good stead. . . She went down and bit the old man. . . " Similar errors also persisted in the subsequent retelling.

Similar defects were noted in his description of a narrative picture: it abounded in disconnected details and—haphazardly developed relationships. For instance, this is how the patient described the famous picture "Another Two": "It looks as if the schoolboy has just come from school... (a long pause)... and his best friend is running to him... and he is upset... and that's all."

Why is he upset? "God knows..." (Reads on the back of the postcard: "The postcard costs 25 kopecks.") *Look at the picture attentively. This is a famous picture called "Another Two." Now think a bit and retell its plot.* "Nothing to think about, he got a two in drawing, that's all." *Why in drawing?* "And in what else?.. The dog is gay, it smiles. Yes... (Reads) Reshetnikov. "Another Two." A long pause, etc.

Equally conspicuous were the defects in the independent plot development. For instance, when asked to make a short composition on the topic "The North" he refused for a long time, saying that the physician had graduated from an institute and he could speak about the North. He wound up by substituting the mentioned sentence "The airplane was flying and the stars were red" for retelling and refused point blank to fulfil the task.

The patient's condition was characterized by markedly increased inactivity, accompanied by an inability to retain a program of operations and the apparent disorder in the selective system of connections, that affected all his behavior. It was against this background that we resumed our experiments in problem solving, using for this the set of problems described above.

The noted symptomatic picture drastically differed from the one observed in all the previous periods.

The previous stages of his disease featured the following symptoms. The patient failed to work out problems that required a more or less complex solution algorithm. He easily substituted an inert stereotype for the necessary deduction process. But the patient could, however, catch up if help was offered to him and reproduced the entire solution process without erring if given the initial element of the solution algorithm. By then the clinical picture grew drastically different. The patient was still able to repeat the statement of a problem correctly but failed not only to distinguish its main question but even to grasp the solution algorithm given to him. The planned process of problem solving was replaced with fragmentary operations disconnected with the statement, irrelevant associations, which undiscriminatingly cropped up in the patient's mind, or inert stereotypes. This *severe loss of preliminary analysis of a problem and the selectivity of established relations* as well as the inability to profit from the prompted cues were the most important features characteristic of the patient's intellectual processes after the partial resection of the left frontal lobe. Hence, whereas prior to the resection of the left frontal lobe, we dealt with a patient who lacked spontaneity but could, however, master the system of operations by distinguishing and retaining the proper algorithm, after the resection the selective nature of his intellectual processes was impaired and the structure of his intellectual activity severely disintegrated.

We shall attempt to illustrate these facts by specific test findings beginning, as before, with the simplest problems and progressing to more difficult ones.

(1) The patient could work out the *simplest tasks* of the type $a + b = x$ or $(a + b) - c = x$ as their solution algorithm is unambiguously determined by the statement and does not require any preliminary reasoning.

For instance, the patient immediately solved the following problem presented to him on the 14th day after the operation: *Fifty seven pine and 33 birch boards were brought to the workshop. How many boards were brought to the workshop altogether?* He said: "57 + 33 = 90. Nothing to solve here, it is already solved."

The process of problem solving for the second, slightly more difficult problem, was similar: *45 kg of oranges and 35 kg of apples were brought to the kindergarten. The children ate 28 kg of the fruit. How much fruit was left?*

The patient repeated the problem correctly. When asked to solve the problem he first said: "I don't know how to solve it." But after the question "In how many operations is this problem?" he answered: "In two operations." and immediately said: "We should first add everything up, that'll make 80 kilograms and then we'll subtract but I don't know what the result will be."

The same problem presented a day later was successfully solved by the patient.

Hence, the patient did not reveal any more or less apparent impairments when dealing with "linear" problems solved in one or two operations. He not only retained the question of the problem but also successfully distinguished its algorithm and performed the respective operations.

(2) It was slightly more difficult for the patient to resolve problems that retained the linear structure but required a number of intermediate operations performed from memory. This group may be exemplified by the problems of the type:

$$a + (a - n) = x$$

To solve such problems we should first derive the value of the second term and only then operate on it, having discarded the earlier performed auxiliary operation.

At first the patient was absolutely helpless in solving such a problem, but managed to work it out after it had been split into consecutive elements.

Here follows the respective illustration.

On the 15th day after the operation, the patient was given the following problem: *A father is 49-years old; he is 18 years older than his son. How old are two of them together?*

The patient repeats the problem correctly, but when asked to work it out he says he can't do it and suggests that the experimenter should solve this problem instead of him.

The patient remembers the statement of the problem but his intellectual process is blocked.

How will you solve this problem? "Maybe I won't solve it at all..."

To facilitate the process of problem solving for him, the problem is split into separate questions. *You should first learn how old the father is now...* "The father!" *How will you learn this?* The patient consults the statement. "He is now... it is said he is 49." *Then should we learn this?* "No, we needn't..." *What should we learn then?* "How old the son is! 49 − 18 = 31!" *Why did you subtract 18 from 49?* "I don't know, my heart prompts it to me."

The patient cannot distinguish the question of the problem and repeats instead the original statement.

The operation is performed but the reasons for it are not formulated.

And what should you do then? "How old are the two of them together. 49 + 31 = 80!"

The problem is solved.

Similar problems were given to the patient a week and two weeks later (on the 22nd and the 31st days after the operation). In both cases they were solved in the same way.

Hence, the patient is confused when dealing with more difficult problems, that require the preliminary determination of one of the terms. He manages to solve such problems only by splitting the statement into separate questions so that each of which unambiguously determines the desired answer.

(3) The solution of such problems does not yet reveal the massive changes that took place in the patient's intellectual processes after the resection of the large section of the left frontal lobe.

These changes grow increasingly more evident at the next stage when he begins to work out problems with a more complicated solution algorithm, which comprises numerous elements and requires the identification of an unambiguous solution method. These are, for instance, problems in which it is necessary to determine two unknown values, deriving them from two equations.

We used to this end a problem which had been already presented to the patient: *One boy bought one pen and one ABC book and paid 37 cents for them; the other boy—two pens and one ABC book and paid 49 cents for them. How much do one ABC book and one pen cost separately?*

It is known that at the previous stage of his disease the patient could not solve this problem on his own and was easily affected by the inert stereotype (the summation of the problem components). He managed, however, to solve this problem successfully as soon as he had been given the initial element of the solution algorithm ("What is the difference accounted for?")

The present picture of his intellectual processes is absolutely different. Even after the patient has repeated the statement correctly, he is unable either to distinguish the respective sequence of operations or catch the given cues. Even after the patient has been given the initial element of the solution algorithm, he does not initiate the required selective system of operations and the solution of the problem, well-secured in his previous experience, is replaced with a series of fragmented attempts, which has nothing to do with its solution algorithm.

Here follows an extract from the official record.

The experiment was conducted on July 9, 1963, on the 12th day after the operation and should be seen in the context of the patient's general disinhibition. The investigator reads aloud the statement of the above problem. The patient repeats the statement correctly but does not begin to work it out.

Solve the problem. "I don't know yet how to do it . . . " *And what would you learn first?* "I don't know... Well... how much does one boy cost..."

The first word of the statement is persevered.

Repeat the problem. "One boy bought a pen and an ABC book for 37 cents, another boy — two pens and two ABC books for 42 cents. What to begin with? With the parts, is that right?"

A number of persevering errors (2–2–42) in the repetition.

Repeat the problem once more. "One boy bought an ABC book and a book... I forgot its title... No, one ABC book and and an ink-stand . . . " *What is the problem on?* "That is the question."

The analysis of solution alternatives is replaced with the scheme "in parts" used before the latest operation.

To facilitate the solution the problem is presented in writing (the scheme of the problem):

1 pen + 1 ABC book = 37 cents
2 pens + 1 ABC book = 49 cents

The uncontrolled recurrence of side-associations. The problem disintegrates.

The patient immediately underlines the equations and writes: 3 pens + 2 ABC books = 86 cents." *Well, and what then?* "That's all!"

We again see the recurring stereotype of summation. The main question of the problem gets lost.

To facilitate the solution, the statement of the problem is presented in writing (the scheme of the problem).

What was the difference in purchases? "In the first case there was one pen, in the other one — two pens." *What can we learn now?* "I don't know what to do..."

The patient notes the difference but does not use this observation for problem solving.

If the difference lies in pens, what can we learn from it? "How much one pen costs . . . But how can I learn this if you give them together?" *Look attentively, you can learn this.* "Probably, I can but the question is how . . . " *Consult the statement. Everything is given there. What are you going to learn?* "How much does one pen cost, is that right?" *That's right.* "Which numbers shall I take — the former or the latter? Well, so... 49... to divide... and what into, I don't know..."

The search for the proper solution algorithm is replaced with fragmented arithmetical operations.

And what would you divide for? "But can it be otherwise? Should I multiply after all?.. I think it is necessary to do the following 49 ÷ 2 = 28." *What will that be?* "49 ÷ 2 = 28. This is the cost of one pen."

The stereotyped repetition of the problem's questions with the loss of the proper denominations and the subsequent correction.

Did you solve the problem? "Not yet... we should learn how much one pen costs... That is, the pen — we already know... 28 cents... now we should learn how much an ABC book costs." (keeps silent)

The stereotyped repetition of the operation of summation.

How will you learn this? "37 cents + 49 cents = 88 cents." *What do you learn by it?* "How much one ABC book costs..." *Do you?* "No... this is already how much 3 ABC books cost." *And how can you learn how much 1 ABC book costs?* "In the same way too... 28 cents... this is the result" *How did you get this result?* "I subtracted 28 cents from 49 and got 25 cents... that is, 21 cents. This is the price of an ABC book."

The contaminations "ABC books" and "pens." The fragmentary operation of subtraction.

Did you solve the problem? "Yes, I did." *Did you solve it correctly?* "Correctly."

The patient does not doubt the correctness of his solution.

It is evident that, in contrast to his problem solving prior to the partial resection of the left frontal lobe, he does not make any attempt to develop the general solution plan or use the first element of the prompted solution algorithm. The true solution is replaced with fragmented numerical operations, each of which has only an indiscriminating relation to the problem statement.

With a view to making it still easier for the patient to work the problem out, it is split into parts and presented to him in the form of a series of separate questions. The following extract from the protocol shows that this split neither helps the patient chart the proper plan (which was well secured in the previous period) nor leads to the development of the required selective operations.

Let us solve it all over again. What do you learn first? "How much does one pen cost."

What will you do for it? "49 ÷ 2..." *Is that correct?* "Correct." *But why should you divide 49 into two?* "This is what is required in the problem. Could it be that I should have divided into three?!"

The question is distinguished correctly but the solution is replaced with the inert stereotype of the direct halving of the sum.

Look at the statement more attentively. Probably you should not divide at all. Think, what accounts for the

The patient works out the first part of the problem

difference in the cost of two purchases? "Just a moment . . . 49 ÷ 37 = 12 . . . one pen costs 12 cents . . . This is too little, therefore, it is wrong." *Why did you do this?* "Well, this very "why?" troubles me a lot. . . Can it be that I should have added?" *No, tell me first how it should* be *done!* "I can't tell you this. . . Should I add, after all?"

but immediately substitutes side-associations for the further solution. He does not understand what should be done to solve the problem and the performed operation is easily replaced with a different one.

What are you going to do then? "Then I'll learn how much an ABC book costs." *What should you do for it?* "Well, this is what I'm absolutely unaware of." *Consult the statement.* "It does not tell me anything at all. . . "

The patient remembers the second question but it does not lead to any attempts to find a solution method.

Repeat the problem again. "And what for? I already know . . . well . . . one ABC book(!) was bought . . . (The rest of the statement is repeated correctly.)

Solve the problem. "49 − 37 = 12. This is the price of one pen." *Explain why you did this.* "It seems to me it should be this way. Is that right or wrong? I don't know. . . Well, first time I divided and now I am subtracting. . . I must have hit the nail if you say this is right!"

The performed operation is reproduced but cannot be accounted for. We see again the stereotyped operation of summation with the persevering in number "88."

The patient is once again presented the statement of the problem this time written in the form of a column.

Will such a statement help you? "I don't know what should be done — should I add or subtract." *Look at the statement attentively and reason aloud.* "Okay. . . There were 2 ABC books. . . and 88 cents for all of it. . . 36. . . 3 pens cost and all in all. . . there were 88. Well, the rest for the ABC books. . . " *What should you learn then?* "We should subtract."

How much did one ABC book cost? "A lot. . . probably, about 20 cents."

Wild guessing instead of reasoning.

The repeated attempts to develop the correct solution of the problem on the 15th and the 22nd days after the operation also failed. In the course of the experiments we once again witnessed similar intellectual processes.

The cited facts prove that the partial resection of the left frontal lobe completely disordered the patient's selective intellectual process, which essentially consisted of the search for the desired algorithm and the discourse in problem solving. Prior to the resection, this process was potentially intact and additional stimulation and the prompting of the initial element of the solution algorithm enabled the patient to initiate problem solving. Whereas after the resection, he was able to perform only disconnected fragmentary operations, which did not form an integrated system and replaced an organized system of intellectual operations.

Since the disintegration of selective activity is already apparent in dealing with the above relatively simple problem, it goes without saying that the disintegration of the program for intellectual processes and their replacement with fragmented operations, are even more pronounced in working out more difficult problems.

We shall illustrate this conclusion by two examples. To this end we shall compare the solution of one and the same problem before and after the partial resection of the left frontal lobe and analyze the impact of training on compensating for the impairment.

(4) We should first consider the process of problem solving in the case of a relatively difficult problem. Its final answer can be obtained only after the preliminary determination of an intermediate value followed by a "reverse" or inverted operation. Let us compare the process of problem solving at the two above stages of the disease.

The patient gets the problem: *A son is five years old. In 15 years his father will be three times older than the son. How old is the father now?*

The experiment of May 22, 1963

The patient repeats the statement of the problem but does not get gown to solving it on his own.

You should now analyze this problem in parts! "And I don't know what the order should be..." *Can you answer the question of the problem at once?* "No, most probably not."

The patient cannot solve the problem unaided.

The patient is given a table of successive questions which make up the solution algorithm of the problem and he is asked to look at the table attentively and answer the respective questions.

What can you learn? "Probably, how old will be the son in 15 years." *And how will you learn this?* "Well, I don't know... maybe 5 x 15 = 45." *Why should we multiply?* "I don't know... probably, this is what we should do."

How old is the son now? "Five-years old." *And how old will he be in 15 years?* "15 + 15... no, 5 + 15 = 20. (Reads the the table further.) Now how old will the father be when the son is 15-years old? 20 × 3 = 60."

The required sequence of operations is performed through consecutive questioning which externalizes the solution algorithm of the problem. The patient

Did you answer the second question? "Not yet." *What did you learn?* "How old the father will be when the son is 20-years old."

is unaware the place of the performed operation within the general solution system.

What should you learn now? (Reads the statement.) *When was the son 20-years old?* "In 15 years." *And how old was the father by then?* "20." *And how old is he now?* "45." *How did you learn this?* "60 − 15 = 45."

The problem is easily solved when the patient is asked separate questions.

The experiment of July 15, 1963

The patient repeats the statement of the problem but its question gets lost." How old is now . . . and who I forgot . . . maybe, the father and, maybe, the son."

The problem is read again. He repeats it correctly.

How are you going to solve it? "I don't know." *What should you learn? How old is . . .* "The mother now . . . no, there is nothing about mamma in the problem."

The recurring irrelevant associations.

The problem is read again. The patient refuses to begin to work it out.

What will you learn now? "How old is the son now . . . but it is clear, five-years old . . . Then let us try to learn how old he will be in 15 years. 15 + 5 = 20-years old." *When will he be 20-years old?* " In 15 years." *What can you learn in the second question?* "How old is the father now . . . 49 years old."

The question of the problem is replaced with the repetition of the statement data.

How did you learn this? "It is said in the problem . . . (The problem is repeated again.) Oh, that's what the problem is like, I confused things . . ." *You have learned how old the son will be in 15 years. What else can we learn?* "From this figure?" *From it and all other data.* "We can learn anything but my age."

The patient perseveres in the number used in the previous problem. Side-associations persist.

What do you learn in the second question? "How old the father is now . . . (The statement is repeated again.) No, I don't know . . . " (Refusal)

The experiment conducted on July 31, 1963

The experimenter once again asks the patient to solve this problem, but this time gives him a table of consecutive questions that promote his orientation of the statement and stipulate the dynamics of the process.

How old is now... "Who?" *Did you answer the first question?* "How should I put it..." *Can you move to the second question?* "I don't know..." *What was the first question?* "How old is the son..." *And what does the statement say?* "Ah...in 15 years. "*How old will he be in 15 years?* "15 x 3 = 45 years."

This illustration shows the great difference in the attempts to work out the problem in both cases.

In the *first* case the externalized algorithm *involves the patient in intellectual activity, helps him retain the results of the previous operation and go over to the following one*, whereas in the *second* case the externalization of the solution algorithm *does not provide any continuous system of operations*, does not help the patient develop any program of operations and, on the contrary, triggers off irrelevant-associations which crop up irrespective of the statement.

The cited extracts from the records graphically exhibit the disintegration of intellectual activity after the partial resection of the left frontal lobe.

Furthermore, they clearly show that neither the independent solution of problems nor training with the externalization of the overall desired solution algorithm could help the patient, with the bilateral lesion of the frontal lobes, initiate the discoursive intellectual process and that he persisted with direct attempts to answer to the problem simply by sorting out numbers.

The last period of our observations over the patient's mental processes was characterized by the disintegration of his systemic intellectual activity, programmed by the statement of a problem.

Conclusions

We have completed the review of the facts obtained during the study of patient U. This case of a slowly growing tumor, located first deep in the right frontal lobe and then spreading to the left frontal lobe, gave us an opportunity to observe the continuous change of intellectual activity. The change was caused both by the growth of the tumor which compressed and destroyed the respective brain structures, and the operations which led first to the resection of the right and then to the partial resection of the left frontal lobes. We can distinguish three stages in the evolution of his intellectual impairment which took place against a background of general changes in the structure of his mental processes.

The *early stage* corresponded to the early stages of the disease when the tumor was limited to the formations located along the median line close to the walls of the ventricles. This stage did not mark any significant changes in the patient's intellectual processes and led only to their minor retardation which should not be, however, regarded as a specific disorder.

The *first and the second stages* corresponded to the growth of a large intracerebral tumor in the right frontal lobe and caused the general dysfunction of the brain's frontal areas. They resulted in increasingly obvious syndrome of inertness and inactivity. Against this background we observed apparent manifestations of the pathological inertness of established stereotypes. It is worthy to note, however, that this pathological condition of the brain did not yet cause the radical disintegration of the programming of complex intellectual processes. The patient, who was unable to chart the solution plan on his own and was easily affected by the developed inert stereotypes, was nevertheless able to use the solution method prompted to him, carry the problem solving process through and compensate (with the support of the externalized elements of the solution algorithm) for the impairments of intellectual processes caused by inactivity and the pathological inertness of established stereotypes.

The *third stage* was connected with the rapid continual growth of the tumor which spread to the left hemisphere and necessitated the partial resection of the left frontal lobe. It brought about radical changes in the earlier observed symptomatic pictures.

These two factors resulted in the disintegration of potentially preserved intellectual processes controlled by a certain program; the selective actualization of intellectual operations was no longer possible; the patient ceased to distinguish the program for problem solving or retain the algorithm given to him. The normal realization of intellectual processes was replaced by irrelevant associations or inert stereotypes, which by then were no longer corrected. Organized searching for proper programs was replaced either with wild guessing, which did not form an integrated system, or with fragmented operations. The patient was no longer able to compensate for the impairment by systematic training.

This case gave us a perfect opportunity to observe all the phases of the gradual disintegration of intellectual activity of *one patient*. It gave us a rare chance to analyze the logic of the aggravating pathological process and develop an insight into the role played by the frontal lobes in securing normal intellectual activity.

---

[1]We do not take into account the decompression trepanation in 1959.

[2]The methods of rehabilitation training used for this patient are discussed in greater detail in Chapter 9.

Chapter 6

IMPAIRMENTS IN PROBLEM SOLVING WITH ATROPHY OF THE LEFT FRONTAL LOBE

We discussed the disorders of problem solving caused by an injury to the basal sections of the frontal lobe and the posterior-frontal sections of the left hemisphere. Then, we thoroughly analyzed the changes resulting from the evolving pathological process which gradually destroyed the normal functioning of the frontal lobes and necessitated first, the resection of the right lobe and then, the partial resection of the left frontal lobe.

Presently we are going to discuss intellectual changes caused by the atrophy of the white matter of the frontal lobes which brought about a severe and stable dysfunction of the frontal cortex.

In the absence of general cerebral manifestations (hypertension-dislocation symptoms) we may disengage ourselves from a number of additional factors which hinder, of course, the review of the obtained data and study in sufficiently pure forms those impairments in the complex types of intellectual activity related to the pathology of the frontal cortical areas.

We shall analyze disorders in problem solving, using as an illustration a case when the growth of a tumor (arachnoendothelioma) in the medial sections of the frontal area resulted in the gradually evolving atrophic process in the white matter of the frontal lobes which we could observe for a sufficient period of time.

Patient B: General Information

Patient B, 47-years-old, a colonel in the air force, was admitted three times to the Institute of Neurosurgery: the first time with a parasagittal tumor of the left frontal lobe, then with the atrophic process in the frontal lobes caused by the continued growth of the tumor.

He fell ill in 1953 when he was 37 years old. The patient suffered from high blood pressure and due to this background gradually developed stiffness in his right leg, then headaches and memory lapses. During an examination in the Riga Hospital, the physician noted papilledema of the optic fundi, anosmia to the left, and hyposmia to the right. The use of contrast methods revealed the dextroposition of the falx. The assessment noted minor manifestations of euphoria and his insufficiently critical attitude to his condition.

At this time he did not undergo surgery but had to retire. In the three subsequent years, the above manifestations gradually grew increasingly pronounced. His headaches grew worse, and the stiffness in the left leg increased and troubled him while walking. Therefore, in 1957 he was admitted to the Institute of Neurosurgery (case history No. 21,178).

The assessment at the Institute noted pronounced pathological manifestations. The patient was euphoric, uncritical, and exhibited apparent intellectual disorders, but believed he was healthy. The percussion of the cranium was painless. We noted stagnant signs in the fundus of his eyes and left-sided anosmia. He suffered from paralysis of upward gaze, hyperesthesia of the right half of the face, static and gait disturbances, Barre's sign to the left, and pathological reflexes to the right.

His behavior also changed a lot—we noted speech disinhibition with the easy recurrence of irrelevant associations and perseverations against a background of general hypodynamia, stiff movements, and the tendency to echopraxia and echolalia—he was apparently insufficiently critical of his condition.

The EEG showed a focus of pathological activity in the left frontal-temporal area against a background of general cortical changes. The ventriculogram showed that the anterior horns of both lateral ventricles were greatly compressed and severely dextrapositioned.

All these facts testified to a large parasagittal tumor in the frontal area of the brain, possibly an arachnoendothelioma which spread more to the left, but caused the dysfunction of both frontal lobes.

In January 1957, the patient was operated on by Prof. Koreisha who removed a large arachnoendothelioma which came from the falciform ligament and extended forward to the frontal lobes of the brain, spread downward to the lower sagittal sinus, and extended backward. The patient was discharged from the hospital in satisfactory condition, but some time later the described symptoms began to gain momentum.

A year later the patient complained of attacks which began with the sensation of an electric shock in his head and forced screams which sounded like reiterated yells "da-da-da" [yes]. Then followed convulsions down his right extremities, bending of the head to the right, and loss of consciousness.

With these symptoms the patient was readmitted to the Institute of Neurosurgery at the end of 1961 (case history No. 32,576) with suspected continued growth of the tumor.

His assessment noted the following picture:

- there were no general brain disorders;
- the stagnant manifestations in the eyeground disappeared;
- his sense of the smell improved;
- we did not note any disorders either of sensitivity or in the motor and reflex spheres.

The apparent psychic disorders, however, persisted and even aggravated. The patient was disinhibited, euphoric, believed that he could resume his work, and noted only slight "lapses of memory" (the detailed data of the psychological assessment is given below).

The arteriogram showed the dextroposition of the anterior cerebral artery, the ventriculogram — measurable hydrocephalus of both lateral ventricles compressed from above.

The findings of the assessment testified to the progressing cicatrizing process (resulting in epileptic attacks) and the atrophy of the matter in the frontal area as well as to the continued growth of the parasagittal tumor in the frontal sections of the brain.

In January 1962, the patient underwent a second operation. Prof. Koreisha found a tumor (arachnoendothelioma) growing from the middle third of the falciform ligament and located deep in the longitudinal fissure. The surgeon removed a tumor 3 x 2 x 4 cm in size.

The postoperative period went smoothly and the patient was discharged without any apparent neurological symptoms. His psychic changes (euphoria, disinhibition, and uncritical attitude), however, persisted. He suffered from behavior disorders throughout the following year. The patient often declared that he was "dead," refused to eat, etc. He would be in high spirits and laugh, and then all of a sudden cry. He believed that he suffered most from the nonexistent "stiffness of the jaws" while his "memory and mind" were, according to him, in order.

His behavior was often impulsive and lacked motivation. The patient, who lived in Dnepropetrovsk, once decided to visit his relatives on business—he took a taxi and asked the driver to take him. . . to the Crimea.[1]

From time to time his euphoria, disinhibition, and uncritical attitude to his condition gave place to languor and inactivity; the patient grew emotionally unstable, and his state bordered on depression. He developed headaches that were even more severe. Against a background of headaches, attacks recommenced which began with indistinct sensations in his arms and legs and were accompanied by the loss of consciousness.

In this state, the patient was admitted for the third time to the Institute of Neurosurgery in February 1963. His optic fundi were normal, his eyesight was normally acute, he did not suffer from either sensory or motor disorders, and he exhibited no pathological reflexes. We could note only a slight hyposmia to the left and infrequent attacks of the described type.

When the patient came to the Institute of Neurosurgery for the third time, he was playful, and laughed inappropriately. He said, smiling gaily, that he had come there "in order to die" and asked to be transferred to the "dead house." At this period the patient was mainly languid and indifferent to everything.

His neuropsychological assessment noted the following picture: like before, the patient had no pronounced gnostic and speech impairments, he easily memorized short word series presented to him and encountered no difficulty in performing simple arithmetic. But the evaluation of his praxis immediately noted apparent disturbances. Imitating the position of hands shown to him, the patient easily slid to the mirror-like reproduction of their position and corrected his mistake only after a special command, but then made the same mistake again. The patient did not show any defects in the test on reciprocal coordination, but whenever he performed a consecutive series of movements he generally simplified the program given to him (for instance, the desired alteration of positions "fist-edge-palm" was replaced by "fist-palm," "fist-palm") and he did not notice his mistakes. In the test on rhythmical beating presented by ear, the patient often added unnecessary, uncorrected strokes while performing the same test after a verbal instruction and he usually slid into the reproduction of established inert stereotypes. He was good at performing the test on selective response (showing his right hand in response to one signal, and his left hand in response to a different one); he easily developed a stereotype and gave inadequate impulsive responses when this stereotype was broken.

The patient easily recited series of four and even five words, but when the test was prolonged he began to inertly reproduce the mastered word order. The patient sometimes assimilated some words (he said "dom - lyes - kol - zvon" [house-forest-stake-peal] or "dom - kol - stol" [house-stake-table] instead of "dom - lyes -kot - stol - zvon" [house-forest-cat-table-peal]) and did not notice his mistakes.

His behavioral disorders were especially pronounced whenever he encountered the need *first to recode* the received information and to control his action by the program derived from such preliminary recoding.

For instance, if given the directions in which a required sequence of operations disagreed with their word sequence, the patient, as a rule, echopraxically performed the operations, reproducing the word order of the verbal directions. He corrected his mistake only after the respective instructions. This did not, however, prevent him from repeating such an echopraxic impulsive action over again. We failed to teach him to inhibit the direct impulse and first to recode the received instructions.

This phenomenon can best be illustrated by the directions to "put the triangle over the cross" or "the cross under the point." Performing this action downward, the patient each time put it echopraxically (without recoding the directions) and persisted with such impulsive actions despite the corrections made.

We also observed severe defects when the patient performed intellectual operations which required the continuous switchover from one operation to another and the elimination of established stereotypes. Thus, while performing serial seven subtraction, the patient experienced great difficulty as he was unable to alter the direction of counting, to turn the obtained difference into the minuend and to continuously overcome the inertness of the preceding number. Hence, the required series took the following form: "100 − 7 = 93...93...97...93 − 7...well, 86...87...81...81...90...70...76...76 − 7...59...59 − 7 = 52..., etc."

It is characteristic that he could not eliminate such uncontrolled impulsive or persevering answers, and only the use of special detailed operations relying on external auxiliary supports allowed him to somehow balance the described defect.

The patient faced similar difficulty in understanding narrative pictures or retelling texts.

His assessment of the idea of the thematic pictures was quite impulsive and limited to distinguishing some (not always essential) detail which the patient interpreted in isolation from other details. This is, for example, how he interpreted the picture "The Skating Rink" which showed a man who had fallen through the ice and a group of people coming to his rescue: "Well, this is a soldier . . . and the mother is seeing her daughter off . . . then an old man. . . ." Here follows his interpretation of the picture "The Broken Window" which showed a guilty boy hiding behind a tree while his victim is trying to catch an innocent guy: "A father is sending his son to school. The technical school . . . the boy is in the uniform," etc. Furthermore, repeated encouragement did not result in the consistent logical analysis of the idea implied in the picture.

Similar impulsiveness was also observed in the retelling of stories read to the patient. He grasped some fragment but did not harmonize it with other components and, therefore, he failed to take in the idea of the story.

The same impulsiveness revealed itself in his filling blanks in the text (Ebbinghaus test) and, finally, in fulfilling assignments on finding logical relationships (for instance, those of analogy) which were most indicative of his inability to inhibit recurring by-associations and subordinate his intellectual activity to a given program.

All these symptoms suggested a *progressing atrophic process in the white matter of the frontal lobes* (especially in the left one) which resulted in a severe dysfunction of the frontal area in the absence of general hypertension-dislocation symptoms.

It was against this background that we carefully studied the structure of the patient's intellectual activity.

Disorders in Problem Solving (The Case of Patient B)

To study patient B's intellectual peculiarities, we offered him a set of problems beginning with the simplest ones whose solution algorithm was easily determined by the statement, and ending with more difficult problems which the patient could solve only by choosing their solution algorithms from a number of alternatives.

(1) Our observations showed that the patient did not encounter any serious difficulty in working out *simple problems* whose solution algorithm was easily determined by the statement.

For instance, he easily solved the problem of the type:

$$a + b = x$$

(*47 pine boards and 35 linden boards were brought to the workshop. 59 of them were used to make benches. How many boards were left?*) and explained correctly what he had done to solve it.

It was equally easy for him to solve a problem on subtraction ($a - b = x$) and a slightly more difficult composite problem of the type: $(a + b) - c = x$. When solving the latter problem the patient repeatedly changed some figures, but persevered the correct solution algorithm (see the following illustration from the protocol of the experiment).

The patient receives the problem: *47 pine boards and 35 linden boards were brought to the workshop. 59 of them were used to make benches. How many boards were left?*

"The pine and the linden — these are trees . . . Then, 82 − 59 . . . " *What are you learning?* "How many were left." *What do you learn in the first question?* "How many boards were brought altogether." *And in the second?* "How many were left. So, 82 − 37 . . . no, 59 . . . that will make 13."

The patient contaminates the numbers. Otherwise, the problem is solved correctly.

Hence, we may conclude that the patient did not face any serious difficulty in resolving simple and composite problems whose solution algorithm was determined by the statement. The patient's pathological peculiarities revealed themselves only in certain pathological inertness of the derived elements.

(2) The patient encountered much more serious difficulty when we offered him simple but *inverted problems* of the type:

$$a + x = b \quad \text{or} \quad x - a = b.$$

The solution of such problems did not require any intricate operations, but their inverted structuring (when the unknown value was mentioned at the beginning or in the middle of the statement) conflicted with the habitual scheme in which the unknown value is always put at the end of a problem and is phrased in the form of a question. In these cases, we noticed that the patient easily performed the required arithmetical operations, but could not repeat the statement of a problem correctly; he changed it in keeping with the secured stereotype and asked the question on the value given in the statement.

Here follows an illustration of his difficulties.

The patient gets the problem: *The boy had 24 apples. He gave some of them away to another boy. Seven apples were left. How many apples did he give to the other boy?*

The patient repeats, "The boy had 24 apples; he gave seven apples away. Seventeen were left."

The problem is read again and the patient is requested to repeat it.

The problem is repeated in the usual and not in the inverted order and immediately solved. All further attempts mark either the

"A boy had 24 apples. He gave some of them away. How many apples were there altogether? (The problem is repeated once more.) A boy had 24 apples... he gave some of them to another boy. How many were left?" *Did you repeat the problem correctly?* "Yes, I did. (The problem is repeated again. The experimentor explains the statement to him.) A boy had 24 apples. He gave some of them away. How many did he give to another boy?"

inert reproduction of the usual stereotype or the contamination of the read statement with the usual stereotype. He does not understand that he repeated the problem incorrectly.

The problem is read in parts. The patient repeats each part correctly. It is, however, still difficult for him to repeat the whole problem. "So... there were 24 apples... he gave an unknown number of apples away... how many apples did he give away?"

We got similar results with other problems of this type. This is how the patient solved the following problem: *There were some birds sitting on a tree. Four more birds joined them. All in all there were now six birds. How many birds were originally sitting on the tree?*

"The birds were sitting. Six more joined them. How many birds were there altogether?"

One of the components of the statement is repeated in the form of the question.

Solve the problem. "Two birds." *Explain, please.* "Since there were all in all six birds and four flew away, then two birds were left."

While solving the inverted problem the patient replaces it with the noninverted one.

The most interesting feature of these research findings is that, in contrast to all the previous cases, patient B not only alters the inverted problem, substituting a noninverted one for it, but also fails to realize that he has changed its question.

It will be shown below that *an inability to retain the question of a problem and an easy substitution of an inerted stereotype for the original statement* are the most characteristic features of the observed intellectual disorders.

The patient's disorders in problem solving are not limited to the unstable retention of a problem's statement and the substitution of a habitual stereotype for a more or less unusual statement. They are even more pronounced in the process of *problem solving* proper.

The patient retains an ability to perform counting operations, but is unable to subordinate them to an accurately perceived statement; he sometimes grasps only a fragment of the statement and slides to specific operations which disconnect with the program for problem solving. But even if he manages to solve the problem correctly, he interprets the obtained results in keeping with the altered stereotyped question which inertly determines their denomination.

Here follows an extract from the protocol which illustrates how the patient was solving a problem identical to the one analyzed above.

The patient gets the problem: *A boy had 24 apples; he gave some of them away to another boy. Seven apples were left. How many apples did he give away?*

The patient repeats the problem: "A boy had 24 apples . . . some of them he gave away . . . seven apples to another boy." *What is asked in the problem?* "How many did he give away. . . How many were left."

While repeating the statement, the patient contaminates the data of the statement with those of the stereotype. The question of the problem is replaced with a component of the statement.

The problem is read again. After the second reading it is repeated correctly.

The problem is again repeated in the usual form.

Solve the problem. "A boy had 24 apples. . . he gave seven away. . . how many were left. . . 24 − 7 = 17." *No, you repeated the statement incorrectly.* (The patient is given the text of the problem to read) "Then, we shall learn how many apples were left." *Did you repeat the question correctly?* "Correctly. . . then how many did he give away. . . 24 ÷ 7." *Why?* "Because there were 24 apples and seven he gave away . . . Seven were left . . . How many did he give away? It is difficult."

He persists in repeating the problem in the habitual form. The disagreement between the repeated variant and the original statement results in attempts to solve the problem by division. The word "part" triggers off the operation of division.

Try to solve the problem, it is easy. "Some part he gave to another boy . . . seven were left. Maybe, 24 ÷ 7. . . and we'll learn how many he gave away." *Why did you divide?* "We won't learn anything otherwise. . . 24 ÷ 7 = 8. . . 7 × 8 = 56. . . 24 ÷ 7 = 4. . . 4 × 7 = 28. . . 3. . . 3 × 7 = . . . 24 ÷ 7. . . 4 × 7 = 28. . . 3. . . 3 × 7 = 21. . . 3. . . three point something. . ." *What does this mean?* "Three apples point something were left." *Why did you divide?* "He gave some part away. . . it is asked here how many he gave away. . . therefore, we should divide."

The patient is offered a table:

_______ Total 24 _______

7 x was given away

_______ left _______

This table graphically shows the correlation between the original sum and the difference. The patient is requested to solve the problem.

With the support of the visual aid (a table) the patient solves the problem correctly.

"You have 24 apples... you gave some of them to me... and you still have seven apples... so, 24 − 7 = 17." *What did you learn?* "How many apples were left." *Read the statement.* "He had 24 apples, he gave some of them away... seven apples were left." *Then what is 17?* "Seventeen apples were left." *Doesn't it contradict the statement?* "Not at all. (Reads the statement again.) No, it doesn't contradict the statement." *Then what do you think, how many apples were left?* "Seventeen." *And according to the statement?* "Seven." *No contradiction here?* "Well, he gave 17 to another boy." *How many were left?* "Seven." *Tell me how you solved the problem.* "24 − 7 = 17." *What did you learn?* "How many apples were left."

We would like to draw your attention to two points in the analysis of the above illustration. On the one hand, even after the patient has repeated the inverted problem correctly, he persists in working out the inertly secured direct problem, and does not notice the contradiction of the denomination of the obtained result with the original statement. On the other hand, he easily diverts from the correct solution algorithm, and having grasped a fragment of the statement "he gave some part away," he begins to divide into parts, which is absolutely out of place here.

Both of these features which testify to a severe impairment of his intellectual process, and which were atypical of the previous cases, grow even more conspicuous when the patient begins to work out more difficult problems.

(3) Whereas the solution of simple but inverted problems marked the inert deformation of a statement which was likened to the well-secured stereotypes of direct problems, the symptomatic picture is much more complicated in working out structurally more complex problems which require several intermediate operations performed from memory. In this case, we could observe *a severe disintegration of overall intellectual activity.* Under such conditions, the inert stereotypes which have nothing to do with the system of the problem may affect its solution algorithm and result both in mistakes in the performance of desired operations and in the substitution of inadequate operations for the correct ones.

Such disorders may be illustrated by the solution of a relatively simple problem from this group: $(a \times b) \div c = x^1$; $(a \times b) - x^1 = x$.

Here follows a respective illustration.

The patient is given the problem: *Some collective farmers had 12 hectares of land. They harvested five tons from each hectare. A third of the harvest they sold to the state. How much was left?*

It is obvious that this is a difficult problem, both because it has a number of denominations [hectares - tons - parts] which should be retained throughout the problem solving process, and because it requires a series of successive operations in which each element begins with the results derived in the previous one; furthermore, the result of each preceding operation turns into the initial data for the subsequent ones.

The patient repeats the problem correctly but easily replaces the final question with an intermediate one.

He can't retain the final question of the problem.

What is asked in the problem? "How much did they sell to the state." *No, it's wrong.* The patient corrects himself and begins to solve the problem. "12 × 5 = 1 = . . . 10 × 50. . . 50. . . 2 × 5 = 20. . . 70. . . was it 70 kilograms or tons? Was it two-thirds or one-third?" *One-third!* "70 − 25 = 65 . . . Why 25? 1/4 — this is 25%."

He begins to solve the problem correctly; the numerical operations are, however, all the time affected by perseverations (2 × 5 = 10. . . 10 × 5 = 50. . . 2 × 5 = 20) and he slides to the bisystems (1/3 = 25 = 25%).

Repeat the problem. "Twelve hectares of land . . . each one yields 5 tons of land. . . 12 × 5 = 70. . . and one-third . . . therefore, 25%, hence, 70 − 25 = 45 kg."

The patient continuously confuses the denominations (ha-ton-kg). Substitution 1/3 = 25% = 25 tons.

No mistake here? "I don't know . . . it seems correct." *Check it out.* "They harvested 70 kg. . . 1/3; 1/3 = 25% . . . 70 − 25 = 45 kg."

He does not understand the defects in his reasoning.

The patient is asked to write the statement of the problem down. He does this correctly and begins to solve the problem: "12 × 5 = 70. . ." *Is it?* "12 × 5. . . certainly, 70 (continues to count, corrects himself) no, 60. . . certainly, it is 60 not 70. . . 60 − 25% . . . 13 = 25%" *Does it?* "One-third . . . one-third . . . how much does one-third make?" *What should we do?* "We should subtract one-third from 60. . . but how much does one-third make?. . . I think, 25 tons." *Really?* "One-third. . . The third part. . . we may divide 60 ÷ 3 = 25. So, 20 kg. . .20 tons. . .60 − 20 = 40."

The mistakes made are inertly persevered; he does not notice them; the mistakes are corrected as a result of additional questions.

The problem is solved.

In this particular case, the patient preserves the solution algorithm of the problem which is relatively simple, although it covers a group of different denominations and a series of successive operations performed from memory. The process of problem solving is impaired in its weakest link — the retention of the respective denominations and the switchover from one operation to another. The pathological inertness of established stereotypes affects the two elements with the ensuing replacement of both the denominations and the arithmetical operations with the persevering stereotypes which, in contrast to the previous patients, he fails to correct.[1] The group of patients to which patient B belongs feature:

- uncontrolled performance of the operations,
- unwillingness to compare the obtained results with the original statement,
- disordered logical reasoning in the course of problem solving.

(4) Similar but even more conspicuous intellectual disorders may be observed in working out a problem which requires operations in parts with the subsequent transition to a semantically comparable but mathematically different problem.

After the patient solved the above inverted problems (see section 2) he was given the following problem to solve: *A housewife used six liters of kerosene in five days. How much kerosene will she use in a month (30 days)?*

The patient listens to the problem and immediately divides. "30 ÷ 6! and that's all!"

The impulsive operation of dividing the available figures instead of the preliminary orientation of the statement.

What are you learning this way? "I am learning how much kerosene was left."

Think, what the question is and how you should solve this problem. "But there is no other alternative, is there?.. It is impossible to solve it otherwise."

The question of the problem is changed and likened to that of the previous problem.

In order to help him remember the statement, the patient is shown a drawing to be used in the course of subsequent problem solving.

The patient neither returns to the statement nor analyzes it. The developed solution algorithm turns into a stereotype. The patient does not analyze whether it is adequate to the problem.

______ 5 d. ______

________________ = 6

____

____________________ = ?

30 d.

Can you calculate using this scheme? The patient steals a glance at the drawing, but does not use it and immediately says, "I can calculate. . . just a minute. . . 30 ÷ 5. . . Thereby, we shall learn how much she uses per month."

The prompted specific problem scheme does not help him analyze it. He persists with attempts to solve it with the help of the impulsive arithmetical operation.

The algorithm of the problem is explained to the patient in greater detail. *This is a problem on parts! Five days equal one part of the month; the housewife uses six liters in this part. How much does she use in a*

The explanation of the solution algorithm does not result in the required discoursive reasoning, but *month?* "Then 6 × 30. . . no, that's wrong." (He keeps silent and does not make any more attempts to solve the problem.)

triggers off another direct operation.

Five days make only one part of a month, a month lasts longer! "Then we'll learn how many such parts there are in a month. . . 30 ÷ 5 = 6. . . 6 × 6 = 36. . . Thirty-six liters."

The problem is solved when the question is overtly presented.

The illustration shows that in this case we witness *the loss of an element which is important for any intellectual activity — that of the preliminary orientation in the statement of a problem.* The patient begins with immediate attempts to work out the problem that has not yet been analyzed and at once initiates fragmented operations which are incompatible with its statement and irrelevant to the denomination of the numbers involved. Even the explanation of the solution algorithm neither enables the patient to analyze the semantic system of the problem nor helps him end impulsive fragmentary attempts. In this respect, this case differs a lot from previous ones. The problem is solved correctly only when the investigator's question formulates the desired operation, which is then easily performed by the patient.

He remembers the explained algorithm and reproduces the mastered solution pattern correctly whenever he is given analogous problems with the identical content but differing numbers. The process of problem solving is impaired in this case only by persevering numbers mentioned in the previous problem. The patient, however, once again faces difficulty when he begins to work out a semantically similar, but structurally different problem which requires another solution algorithm.

After the repetition of the analyzed problem the patient is offered a new one: *A housewife uses ten liters of milk for five days. How much does she use in a week?*

The patient listens to the problem and writes down "How much in seven days" and begins to solve, "If ten liters in five days, then in two days. . . in two more days she uses . . . in five days — ten liters and in seven days . . . I know that that there are two more days. How much then in one day?"

The patient writes the correct question necessary to solve the problem, but he does not initiate the operation.

What should you learn? "How much kerosene does the housewife use in one day." *And then?* "If I learn how much she uses in one day . . . I'll learn how much in two days and I'll add . . . so, it's ten liters in five days and in one day. . . 10 . . . ten minus seven, probably . . . I think, seven . . . three liters."*What did you learn?* "How much in one day. . . and I need in two days. . . then six liters. . . the total of 16 liters."

The patient charts the correct solution plan, but the subsequent operations are actualized irrespective of the charted plan. The result obtained from the impulsive operation is later involved in the solution algorithm.

Check it out. "Ten liters in five days. How much in seven days? We should learn in one day... how much should that be? Ten liters in five days... in seven days... no, that's wrong... it won't make three liters, it'll be less. Ten liters in five days... six liters in two days and I got the answer of three liters in one day... it should be this way..." *Which way?* "Just a minute, we should 16... in five days... the only thing we should learn is... how much milk does she use per day..."

The patient tries to verify the obtained result, but relies in his assessment on the direct impression of the obtained amount rather than the comparison of the performed operation with the solution algorithm.

What should we do for it? "It is wrong to subtract seven... probably we should 10 ÷ 5... then two. But two is too little! Ten liters in five days, in seven days much... uses four liters... in one day... how can I learn this?" *You put the question correctly, but the operation is wrong.* "Then what should I do?"

Attempting to find another solution alternative, the patient randomly performs the correct operation, but he does not compare it with the solution algorithm either.

This protocol highlights the very important intellectual peculiarities of this patient. Even after he has formulated a solution plan correctly, he is still unable to adhere to it while performing the subsequent operations. The charted plan does not determine his further actions and remains disconnected from them. All the patient's subsequent attempts at problem solving are limited to a number of impulsive and fragmented numerical attempts detached from the solution algorithm.

This conclusion can be illustrated by the experiments with the same patient which have been described in a different monograph (see: Luria, 1963, pp.372–374).

The patient gets the problem: *There were 18 books on two bookshelves but they were unequally divided. There were twice as many books on one of them. How many books were there on each bookshelf?*

The patient repeats the problem correctly and begins to solve it. "There are 18 books on two bookshelves... on one of them — twice as many as on the other... no, I can't... if they were equally distributed that would make nine ..." *But there are twice as many on one of them.* "There are nine on one of them... but then that would be equal... then 9 ÷ 2 = 4.5... so, it's 4.5 and 12.5... no, there are four on one shelf and 14 on the other." *Is this correct?* "I think it is." *And where did you get four from?* "Well, if it is not 4.5 then it must be four... a half does not fit here... if it were 20, I would calculate."

The problem is solved by fragmentary operations with the subsequent discarding of the fraction and "leveling" of the result.

The patient does not realize the incorrectness of the operation.

The investigator explains to the patient the method "in parts" for solving this problem. The patient masters its solution algorithm and later uses it for solving identical problems. But when given a similar problem with other figures his difficulties, however, reappear. The used method of tripartition inertly persists and hinders the correct solution, which is replaced with a number of fragmented operations.

The patient gets the problem: *There were 24 kilograms of sugar in two boxes. There was three times more sugar in one of them. How much sugar was there in each box?*

The patient immediately begins to solve the problem. "24 × 3 = 12... we are writing two... four times more... no, 24 × 3...3 × 4 = 12... we are writing two and we shall remember one... four times more... four... nine... 92 kilograms." (The problem is repeated.) "Oh... it's three times more than in the other one. Then it is eight in one box and 16 in the other. 16 + 8 = 24. (The problem is repeated once more.) There are 24 kilograms in two boxes... in one of them there are four more... one-quarter or three-quarters more..." *Is this so?* "No, it's by one-quarter." *It is three times more!* "24... three times more... 24... 48... 48 + 24 = 52... there were 52 in one box... but that's wrong... there were 24 in one box."

The consistent solution is replaced with a number of fragmented operations which are disconnected with the statement.

The repetition of the statement leads to the recurrence of the inert stereotype of "tripartition."

When the patient is told that the problem has been solved incorrectly, he tries to divide into four parts. This attempt is, however, soon replaced by attempts to operate in three parts.

Due to the repeated solution of similar problems, the patient mastered the scheme of dividing into parts with their subsequent distribution. But the patient faced difficulty in applying this scheme whenever he had to deal with new figures, and therefore, to substitute a new ratio for the stereotyped "tripartition." But he encountered the most difficulty if he was offered a problem which looked like the mastered type of problems in parts, but was virtually a problem in difference.

Here follow the respective illustrations.

The patient receives the problem: *There were 24 books on two bookshelves but there were two less books on one of them than on the other. How many books were there on each shelf?*

The patient immediately begins to solve the problem. "24... two less books... then it's 24 ÷ 3 = 8... 8 and 16." *Did you solve the problem correctly?* "24 − 8 = 16... 16 + 8 = 24... Precisely!"

But there were two less books on the bookshelf! "Two less books ... so, it's in parts, too ... in three parts... 24 ÷ 3 = 8... 8 and 16!" *No, you solved the*

The solution of the new problem is replaced with reproduction of the old algorithm. The patient does not realize that it is wrong to apply this algorithm here.

problem incorrectly. There are two less books on the second shelf. What should be done? "We must divide 24 into three...or into two...there are two books less here...then it will be equal."

The algorithm of "tripartition" perseveres blocking the search for a different solution algorithm.

It is noteworthy that when the patient masters the solution algorithm for the second type of problem through special training (Discard the difference; Halve the result; Add the discarded difference to one of the halves) he is incapable of switching over to solve the problems of the first type (problems in parts) due to the inert effect of the second algorithm.

Here follows an illustration.

The patient gets the problem: *There are 12 kg of sweets in two boxes. There are twice the sweets in one of them than in the other. How many kg of sweets are there in each box?*

The patient repeats the problem and immediately begins to solve it. "12 − 2 = 10; there are five in one and seven in the other." *Is this correct?* "Sure it is! Precisely! (The problem is read again.) Twice more . . . Oh, it's not two more but twice more here...(He draws the schema of two boxes.) There are 12 kg in two boxes... there are twice more in one of them... twice more. This does not exceed 12 kg."

The inert reproduction of the mastered algorithm of the problems "in difference."

Despite the reproduction of the statement ("twice more") the patient repeats the inert operation of addition even after he has distinguished the parts.

How can we solve it? "This is precisely the question. I don't know. . . .there are twelve in the first box...no, 2 + 2 = 4...to add 4...12 + 4 = 16... no, that's wrong..." (Refusal.)

This illustration allows us to proceed with the analysis of the patient's intellectual peculiarities. The patient is given a difficult problem in parts. The correct reproduction of its semantic scheme does not determine his subsequent operations and consistent logical reasoning is replaced with fragmented operations disconnected from the problem, which draw him beyond the frame of its statement. After the patient has been trained in using the respective algorithm, which is relatively accessible to him, the mastered solution algorithm turns into a stereotype. This stereotype is inertly reproduced in working out each following problem, and therefore, the patient cannot deviate from the mastered solution algorithm and solve differing problems.

The specific feature of the patient's pathological intellectual activity is that the mastered solution algorithm turns into *an inertly reproduced stereotype*. Beyond its frame the patient can do nothing but perform fragmentary, impulsive numerical operations, undetermined by the semantic system of a problem.

(5) His inability to analyze the statement of a problem, distinguish its solution algorithm, and adhere to the charted program grows even more conspicuous when the patient tries to solve problems which require the comparison of their components and the option of a required solution algorithm.

As in the above cases, these difficulties were especially characteristic of problems with two equations (of the type: $x + y = a; x + 2y = b$). The process of their solution revealed most graphically all the above specific disorders of the patient's intellectual activity.

The patient gets the problem: *A pupil bought one ABC book and one pen and paid 37 cents for his purchase. Another pupil bought one ABC book and two pens and paid 49 cents for them. How much do one ABC book and one pen cost separately?*

This problem is presented to him in writing in the form of a column; its idea is explained.

In contrast to previous cases, its solution takes the form of disconnected fragmentary attempts, in no way determined by the statement.

The patient reads the statement and immediately begins to solve the problem. "First, we shall learn how much one pen costs. 13.5 cents." *How did you get this?* "It's quite simple: 37 ÷ 2 = 23.5 cents. Third, how much do two pens cost? 13.5 + 13.5 = 27 cents. Fourth, how much does one ABC book cost? 23.5 cents."

The systemic problem solving is replaced with the fragmentary operations of halving the figures given in the statement with the persevering changes of numbers. We should note the instability of questions and the persevering return to the already "answered" questions.

Did you solve the problem correctly? "Both correctly and incorrectly." *Why correctly and why incorrectly?* "Correctly because I found how much a pen and an ABC book cost. Incorrectly because a pen costs 12 cents." *Why was it originally 13.5 cents and now it is 12 cents?* "Both figures are correct. But generally it costs 12 cents. And an ABC book? It costs 49 − 12... and the problem is solved."

The question whether the solution is correct does not help the patient compare the statement of the problem with the answer which is out of place here.

And where did you get 12 from? "I contrived it. I was thinking the problem out and decided that the pen should cost nothing but 12 cents." *Write it down once more and solve the problem.* "Here you are... how much does one pen cost? 37 − 12... how much does one pen cost... and the problem is solved!"

The patient can perform some operations correctly, but he quickly loses them or replaces them with the stereotyped operation.

We obtained similar data in the experiments when the patient was solving another problem which required the comparison of several values and the successive derivation of each value.

The patient is given the problem: *Three boys found 87 mushrooms. The first boy found 42 mushrooms, the second boy — 13 mushrooms less. How many mushrooms did the third boy find?*

The patient is requested to repeat the statement of the problem. The first time he says, "Three boys found 48 mushrooms.(He is corrected: 87 mushrooms.)The third — 12 mushrooms less. How many mushrooms did they find together? (The problem is read out loud again.) Three boys found... 87 mushrooms. The second — 42 mushrooms less. We should learn how many mushrooms the third boy found. (The problem is read out loud once more.) Three boys found 87 mushrooms. The first one — by 42 mushrooms... No, 42 mushrooms! The second boy 12 mushrooms less. How many mushrooms did they find altogether?"

The incomplete repetition of the statement; the original statement is repeated in the form of the question.

The problem is once again repeated incompletely. The statement is repeated correctly, but the question is replaced with the reproduction of the value mentioned in the original statement.

Hence, the main impairments of the patient's thinking are apparent already at the stage of statement repetition. The statement is repeated in fragments, and the data of the statement does not make up the problem. But even if the patient manages (after several attempts) to repeat the data of the statement correctly, the question of the problem gets lost, and therefore, he puts the question instead to one of the components of the original statement.

Therefore, *the patient does not grasp the system of relationships which make up a problem, but easily replaces it with an isolated reproduction of data which is not integrated by a common question and does not form the semantic structure of a problem.*

In order to help him retain the statement, the patient is given a written scheme which should serve as a visual support for problem solving.

87 mushrooms	1st — 42 mushrooms
	2nd — 12 mushrooms less
	3rd — ?

Relying on this scheme, the patient repeats the statement and the question of the problem correctly.

Solve the problem. "The first question is how many mushrooms the third boy found . . . 87 ÷ 12. . . that'll make six . . . no, seven . . . no, I can't divide without a remainder . . . therefore, it is wrong." *Why are you dividing?* "I am dividing into the boy (points: "Twelve mushrooms less"), as he found less..." *No, that's wrong!* "Then 42 ÷ 12 = 3. . . No, it does not divide without a remainder either..."

The patient says the final question, ignoring the statement and tries to perform a number of impulsive operations triggered by the fragmentarily distinguished link "twelve mushrooms less."

Consult the problem and think—should you try to learn at once how many mushrooms the third boy found? "Everything is known here but the third

He disconnects the value "twelve mushrooms less" with the rest of the boy!" *Say what is known to you*. "All of them found 87 mushrooms. The first boy — 42 mushrooms. The second one — 12 mushrooms less. . . Oh, yes. . . This is what we don't know. . . then we should learn how many mushrooms the second boy found. 87 ÷ 12. . . " *Why are you dividing?* "And what should I do? Should I add? Okay, I'll try to add — 42 + 12. . . that'll make 54! Then the second boy found 52 mushrooms. 54 ÷ 12. . . no, it does not divide without a remainder . . . then it's also wrong."

statement and takes it in as an absolute number. The specific request to enumerate what is known leads to distinguishing the significant relation: the patient grasps the intermediate question, but he attempts to answer it only fragmentarily.

The continued attempts to solve the problem take the form of analogous fragmentary guesses and end in failure.

This illustration shows that as the patient carries out absolutely no preliminary analogues of the statement of the problem, he is unable to grasp its complex semantic structure; the constituent components of the statement are repeated in isolation, the final question which integrates all its elements into a single semantic entity is lost, and replaced with one of the values mentioned in the original statement.

The preliminary analysis is not carried out even when the statement of the problem is presented *in writing;* while reading the presented statement the patient does not analyze the correlation of its constituent elements which he fails to integrate into a semantic structure. Therefore, it is only natural that the patient neither finds the respective solution algorithm, nor formulates a certain solution strategy, nor tries to develop the respective tactics. He limits himself instead to impulsive operations prompted by the isolated fragments of the problem. It also goes without saying that such a structure of problem solving rules out the possibility of comparing the performed operations and the obtained results with the original statement. Therefore, it inhibits the critical attitude to one's actions as well as blocks the realization and correction of the mistakes made.

(6) In the analyzed case, the presentation of a problem with a large number of unknown elements which required determination of a solution algorithm comprising a series of consecutive operations, resulted in disintegration of the intellectual process, and its replacement with fragmented attempts outside the problem system.

We have, however, already noted that the patient also suffers from a different type of intellectual disorder. Let us now elaborate this point in greater detail.

The patient encounters additional difficulty in working out problems which require two alternating successive operations — addition, providing the earlier unknown value, and division, showing how many times the derived value is greater than the original one. (For instance, a problem of the following type: *A ruler is 29 cm long and another one is 40 cm longer. How many times longer is the second ruler than the first one?*) The point is that in solving such problems the patient is tempted by two detrimental tendencies: the tendency to treat the unknown value of the second component as an absolute figure (to replace "40 cm longer" with "40 cm long") and the tendency to directly divide the second number into the first one (40 cm ÷ 20 cm = 2 times). It is nearly impossible for a patient with a massive lesion of the frontal lobes to overcome the tendency of simplifying the solution algorithm of a problem and solving it in one simple operation. Therefore, such patients face especially grave difficulty in working out conflicting problems of the above type, while the process of solving such problems is particularly symptomatic of their tendency to substitute a simplified stereotype for a complex solution algorithm.

While analyzing such problems we shall first illustrate the process of their solution and demonstrate how the above difficulties can be eliminated; then we shall discuss some special conditions which make them inaccessible to this group of patients.

The patient receives the problem: *There are three liters of milk in a jug; there are six more liters in another one. How many times more milk is there in the second jug?*

The patient reads the problem aloud and immediately says, "Two times! 6 ÷ 3 = 2." *Read the problem attentively.* (He reads.) "It is two times all the same."

The problem is solved in the direct operation of division.

The problem is repeated again and split into stages. It is stressed that the second jug holds six liters more. "Then it is nine liters. That's all!"

Is it? Mind the question, please. "No... it is... how many times here." *Do you know how many liters the second jug holds?* "Six liters more." *How many altogether?* "6 + 3 = 9... So, 9 ÷ 3 = 3! Three times!"

The intermediate operation replaces the direct one. The final question of the problem gets lost. The problem is solved correctly only after it has been split into separate elements.

The patient reveals a tendency to regard a relative value as an absolute one. He substitutes a simple solution algorithm for a complex one, but successfully solves the problem after it has been split into a series of consecutive questions each of which can be answered only in one way.

But this does not at all mean that he has grasped *the entire system of the problem*; therefore, we gave him another analogous problem immediately afterwards to see if this was true.

The patient was offered the problem: *One path is 15 meters long and another one is 45 meters longer. How many times longer is the second path?*

The patient repeats the statement but omits the main question. After the problem is read again, he manages to repeat it correctly. He receives as a visual aid a drawing which graphically shows the length of both paths.

Solve the problem. The patient consults the drawing and says, "15 + 45 = 60!" *What is this?* "This is

The visual aid prompts the operation of addition

the second path longer than the first one." *What should we do next?* "We should learn the length of both paths."

We once again read the question to the patient. *How many times is the second path longer than the first one?* "45 ÷ 15 = 3...Three times." *And what is 60?* "The second path is 60 meters longer than the first one." *Then how many times longer is the second path than the first one?* "45 ÷ 15 = 3...Three times!"

to the patient. But the result is misinterpreted. We see the recurrence of the tendency to add both figures. The answer to the question triggers the direct operation of dividing the original numbers.

Therefore, neither the split of the problem into successive questions, nor the use of auxiliary visual schemata can eliminate the essential defect; the patient is still incapable either of synthesizing the components which form the problem's statement or developing the required solution algorithm. Instead he repeatedly performs disconnected operations which do not form an integrated system, but are determined by the directly perceived fragments of the problem.

It has already been noted that the split of this problem into constituent elements and the use of visual aids allowed the patient to get an insight into its statement and solve it (although the developed skill was unstable). One minor change in the statement was, however, enough to make such types of problems absolutely impossible for the patient to solve. This seemingly insignificant change had to do with the *semantic aspect* of the problem.

In all the above cases, we dealt with the quantitative comparison of two similar objects: we compared the lengths of the paths, the number of liters in the jugs, etc. Therefore, it was relatively easy for the patient to disregard the specific features of both objects (the first or the second path, the first or the second jug) and concentrate on the operation of comparing the respective values. The situation is different if the statement of a problem discusses *two different objects* which, furthermore, differ in size. In this case the value (for instance, the length) of the second object comprises the length of the first one plus some additional increment. Therefore, to solve the problem, it is imperative to disengage oneself from the specific image of the first object and realize that the value of the second object consists of an abstract value of the first one with the respective addition. A person with normal intellectual processes encounters no serious difficulty when performing such an operation with three elements (two objects and their values) or dealing with a comprehensive process. Whereas for a patient with a marked frontal syndrome and pathological inertness of stereotypes, this operation is impossible. It adds another insurmountable obstacle to the earlier faced difficulties in working out conflicting problems.

We shall limit ourselves to one illustration: The following problem is one that the patient failed to solve despite prolonged training.

The patient receives the problem: *A pencil is 16 cm long; its shadow is 48 cm longer than the pencil. How many times longer is the shadow than the pencil?*

It is self-evident that in this case, the patient deals with three different items—a pencil, its shadow, and their length. Furthermore, the length of the second object (shadow) comprises the length of the first one (pencil). To successfully solve the problem it is necessary to disengage oneself from the specific image (of the pencil) and operate exclusively on one feature common to the two objects (their length). This is an additional obstacle to solving the problem.

Here follows an illustration of the patient's process of problem solving. The problem was read three times. The patient could not repeat it accurately enough: the elements of the problem (the pencil, the shadow and the length) and its operations (48 cm longer — how many times longer) did not fit the required system.

To make it easier for him to solve the problem the investigator draws the following schema:

________ 16 cm

the shadow is 48 cm longer

____

____

Next, he asks the patient to repeat and then to solve the problem. The patient glances at the schema and immediately says, "Aha, it's clear. 48 ÷ 16." *What do you learn this way?* "How much longer is the shadow than the pencil." *Look at the schema.* "Aha. We should learn how long the shadow is altogether . . . 48 + 16." *What did you learn?* "The length of the shadow." *And what is asked in the problem?* "How many times longer is the shadow than the pencil. . . 48 ÷ 16 = 3!"

An attempt at direct solution. Using the schema the patient performs an operation of summation.

This operation is not, however, used for further solution and the patient again performs the direct operation of dividing the values given in the statement.

Numerous continued attempts failed to yield the desired result as it turned out that even after the patient had repeated the detailed solution algorithm prompted by the investigator, he was unable to retain it. He repeatedly attempted to solve the problem by uncoordinated numerical operations, as he was unable to disengage himself from the specific images of the objects and integrate the three basic components of the problem into a coordinated system.

Our research findings show that all patients with a marked frontal syndrome are unable to work out this sort of problem and fail to master its solution algorithm even after prolonged training.

Conclusions

The provided data reveals a drastic difference between the intellectual disorders of patient B suffering from massive atrophy of the frontal lobes, and the intellectual defects caused by an injury both to the basal sections of the frontal lobe and the posterior-frontal sections of the left hemisphere.

The previous patients with an injury to the basal areas of the left frontal lobe or posterior-frontal sections of the left hemisphere enjoyed intact intellectual activity which was still determined by the statement of a problem and retained its selective character. Their intellectual operations were impaired as a result of pathologically weak inhibition processes in the former case and the pathological inertness of established stereotypes in the latter. The two phenomena could merge. They were,most probably, caused by one factor, but the pathological impulsiveness or the pathological inertness of established stereotypes could prevail depending on the patient's condition. The statement of a problem, however, always retained its determining effect and a minor additional stimulus was enough for the patient to overcome the deviations from the charted plan.

The symptoms observed in the patient with the massive atrophy of the frontal lobes is quite different.

He, as a rule, did not carry out any preliminary analysis of the problem's statement; its data was not compared and he did not develop an adequate solution algorithm. Therefore, he could solve correctly only the simplest problems when the statement clearly determined the desired operations.

But the patient was unable to work out more difficult problems as their statement was fraught with numerous alternatives, and the proper solution algorithm could be determined only due to its preliminary analysis, the comparison of the constituent elements, and the development of the general solution scheme. A system of consistent operations was replaced with impulsive attempts determined by some fragments of the statement. These were most frequently irrelevant formal arithmetical operations which had nothing to do with the semantic system of the statement as a whole.

These operations were generally unselective in character and they grew particularly chaotic if the statement of a problem comprised a large number of constituent elements and required a hierarchy of operations performed from memory.

It is noteworthy that the patient did not at all compare the performed operations with the statement of a problem. The verification process was impossible for the patient not only because of his poor attention and insufficiently purposeful activity, but also as a result of defective verification means—he suffered from impaired comparison, generalization, and abstraction. The patient could neither notice his mistakes nor correct them.

In summary, the distinctive feature of disorders associated with the massive atrophy of the matter of the frontal lobes was *disintegration of the selective system of intellectual operations.* This disintegration of intellectual activity revealed itself in the complete loss of preliminary analysis of the statement, an inability to draw up the solution plan, to perform operations necessitated by it, and carry through the subsequent comparison of the obtained result with the original statement.

---

[1]Dnepropetrovsk is situated in the Ukraine, a thousand kilometers away from the Crimea. Translator's note.

[2]This problem was given to patient B twice in 1961 and 1963. The process of problem solving was similar in both cases.

Chapter 7

DISORDERS OF PROBLEM SOLVING WITH AN INDISTINCT FRONTAL LOBE

Thus far we have been analyzing intellectual disorders of patients with a massive injury to the frontal lobes and a marked frontal syndrome. Equally interesting are, however, the cases when a lesion of the frontal lobes causes only an indistinct frontal syndrome or sometimes virtually no higher cortical dysfunctions at all. Our concern for the analysis of such intellectual disorders is particularly great as changes in the intellectual processes of such patients are often virtually the only basis for the provisional diagnosis of their lesions.

A careful neuropsychological examination shows that intellectual impairments in such patients reveal the same features which are exhibited in a more conspicuous form by patients with a marked frontal syndrome.

We are going to discuss two typical cases which will allow us to describe in detail the structure of intellectual impairments of an indistinct frontal syndrome.

Patient U, 39-years-old (case history No. 38,243), an engineer, was admitted to the Institute of Neurosurgery in March 1964, complaining of headaches which began in 1963. During the summer of 1963, she had an attack unaccompanied by cramps which resulted in loss of consciousness.

The patient was admitted to the Institute of Neurosurgery with suspected cerebral growth. Her assessment noted papilledema in the fundi of both eyes and a minor decrease of vision down to 0.6; her visual field remained normal.

We observed very few neurological symptoms:

- the patient had no impairments either in sensation or motor functions;
- we could observe occasional equivocal episodes of hyposensitivity in the right cheek and hypertension to the right.

She had a flat nasolabial fold. We did not note any pareses or pathological reflexes. The cerebrospinal pressure was 240; we did not find any deviations in the liquid contents.

Her EEG showed a distinct focus of pathological slow waves in the left frontal-parietal area.

The roentgenogram revealed a vascular conglomeration in the frontal-temporal area of the left hemisphere.

Prof. Rapoport's inference said that paresthesia in the right part of the face testified to the lesion of the opercular area and suggested a tumor which could be located in the area of the fissure of Sylvius; the indistinctness of the symptoms could be accounted for by a well-developed venous system which rendered it possible to balance the blood outflow.

The patient's neuropsychological assessment noted very few symptoms. She was perfectly oriented and intellectually intact. The patient had no gnostic, praxis, or speech dysfunctions. We did not note any disorders in her activity or any manifestations of euphoria. She was adequately critical of her condition. The patient easily performed calculations and showed no impairments in tests on the formal logical thinking.

The only symptom that drew our attention during her neuropsychological assessment was impulsiveness while performing the tests which was, as a rule, corrected by the patient herself. For instance, in the tests on position praxis, the patient first reproduced the shown position in a mirror-like fashion, but then corrected her mistake. Similar defects were revealed in Head's test on the reproduction of the position of the physician's hands. Analogous disorders were observed in the tests when the patient had to draw two spatially coordinated figures (for example, the cross under the circle, etc.). In these cases, the patient often failed to execute the required recoding and performed the operation according to the word order of the instruction. The defect was corrected by the patient only in the course of the subsequent check-ups.

All these tests pointed to a stable impairment in the regulation of her mental activity which revealed itself in the tendency towards a weakened orientation for intellectual activity and premature impulsive reactions.

We could also note another distinctive feature in her auditory-verbal sphere: the patient encountered serious difficulty in memorizing a series of four to five words presented to her by ear. She continuously changed their word order and occasionally repeated developed stereotypes.

Both these symptoms were very indistinct, but their stability and recurrence in many different tests testified to their regular nature and suggested a dysfunction of the frontal sections, affecting the left temporal area.

On April 17, 1964, the patient was operated on by Prof. Egorov who removed from the left frontal-temporal area, a large extracerebral tumor (an arachnoendothelioma) located deep in the fissure of Sylvius, and growing inward towards the medial-basal sections of the frontal lobe.

The postoperative period went smoothly and six weeks later the patient was discharged from the hospital in good condition.

The study of the problem solving process of this exceptionally intact patient gave unexpected results.

The symptoms of impaired orientation for intellectual activity and premature impulsive responses (which were hardly noticeable in the assessment of her gnosis, praxis, and speech) became quite apparent in her problem solving operations. This case afforded us an opportunity to observe the *disintegration of selective intellectual processes* which had been described earlier in its more severe forms for patients with massive injury to the frontal lobes.

The impaired selectivity of the patient's intellectual activity first revealed itself in the assessment of her constructive thinking. While performing tests involving the construction of geometrical designs (Kohs's tests) she replaced the logical analysis of a model with impulsive constructions of individual structural elements which did not help her make the required decision. The same symptoms were noted on tests of the comprehension of written material and its logical development and plan.

But these impairments were most conspicuous when she was asked to solve arithmetic problems.

The patient could easily solve simple arithmetic problems of the type:

$$a + b = x \quad \text{or} \quad a + (a + b) = x.$$

She was equally good at solving slightly more difficult problems which required the preliminary operation of distinguishing parts then followed by multiplication. For instance, *A housewife uses six liters of milk in five days; how much milk will she use in 30 days?* But as soon as she was given a problem which necessitated preliminary analysis of the statement and distinguishing a cycle of consecutive auxiliary operations, she immediately revealed a marked impairment of the orientation basis for the operation and quickly began exhibiting impulsive operations.

In the experiment conducted prior to surgery, the patient was given the following problem: *Three boys caught 11 kg of fish; the first and the second boys caught seven kg, the second and the third boys — six kilograms. How many kg of fish did each boy catch separately?*

This problem, which is fraught with difficulties in the case of a direct solution, can easily be resolved by distinguishing the elements needed in the second and the third equations, and referring the difference to this link. The patient easily repeats the statement of the problem, but refuses to solve it.

To find the proper solution algorithm, the problem is presented to her in the form of three equations:

1st, 2nd, and 3rd = 11 kg;
1st and 2nd = 7 kg;
2nd and 3rd = 6 kg.

The patient looks at the equations and says, "No, I don't see anything here... I don't know how to solve it." She is given further assistance and the problem is written down in the following form:

She neither analyzes the statement nor compares the two latter equations with the first one. She does not distinguish the lacking elements either.

1st + 2nd + 3rd = 11 kg;
1st + 2nd = 7 kg;
2nd + 3rd = 6 kg.

The patient immediately says, "Well, then 11 ÷ 6... that'll make 1.5 kg. . ." *Where did you get six*

An impulsive solution of the equation; the meaning

from? "1 + 2 + 3 = 6 . . ." *What does number six stand for?* "Six boys."

No, there were three boys, these are simply ordinal numbers: 1st, 2nd, 3rd. "Then these are six parts . . . 11 ÷ 6 . . ."

Okay, and what are you going to do next? "I'll multiply 1.5 by 1 — that's for the 1st boy, 1.5 by 2 — that's for the 2nd boy and I'll multiply 1.5 by 3 — that's for the 3rd boy . . . no, there is something wrong here . . . I must have done something wrong."

Yes, you have been solving it incorrectly. Don't take the parts. This isn't a problem in parts. There are three boys here. "Then 11 ÷ 3 . . . then that'll mean that they all caught equal amounts . . ."

No, they did not catch equal amounts. Recall the statement of the problem! "Well, then we can't divide. Then let us assume the first boy caught three kilograms, the second one — four . . ." *Well, and what then?* "I don't know . . ." *You should not assume but learn the precise catch!* "I don't know how to do this . . ."

The patient is given additional help. *Look at the equation attentively. What is the second equation lacking and what can we learn from it?* "I'll subtract seven from 11 . . . that means that the third boy caught four kg." *What then?* "Then we should learn how much fish the third boy caught . . . Probably, we should subtract four from 11 — that'll make seven. . . no, there is something wrong, we are getting seven again."

Now look at the second and the third equations and compare them. "Yes, then it's like this: 2 + 3 = 6 kg . . . Hence, the second boy caught 11 − 4 kg . . . no, 11 − 6 kg . . ."

Have a better look at the equations, is this so? (Looks at the third equation) "Then, 6 − 4 = 2 . . . caught the second and the third we should add and subtract . . ." (The patient refused to solve the problem further.)

of each component is ignored; the ordinal number is taken for the summand; the operation is performed irrespective of the problem's statement. Later the patient inertly continues to manipulate the obtained data, although she is told that it is wrong. She ignores both its inaccuracy (11 ÷ 6 = 1.5) and its disagreement with the statement of the problem. The operation with parts reproduces the scheme of the problem in parts, which was worked out the day before.

Only later does the patient realize that the operation is meaningless.

Another impulsive operation.

When her mistake is shown to her, the patient impulsively goes over to the "assumption" which has nothing to do with the statement of the problem.

The patient solves the problem after she has been asked the leading questions but her further attempts to solve the problem result in impulsive meaningless operations.

She returns again to the mechanical operation of summation; the ordinal numbers are once again treated as the summands.

The nature of the patient's intellectual activity revealed by her solution of this particular problem is quite demonstrative. It turns out that the patient is unable to inhibit her impulsively arising attempts at problem solving and concentrate on the analysis of the statement. The orientation stage of the intellectual operation is altogether lost. A logical process of problem solving is replaced with impulsive numerical operations which do not correspond to the context of the problem. It is characteristic that the individual values of the problem are absolutely meaningless in these impulsive attempts; an ordinal number is treated as a summand and the patient gets a sum in kilograms, although she has added the boys' ordinal numbers. Having noticed her mistake, she rejects the logical analysis of solution algorithms and impulsively replaces it with the arbitrary "assumptions" that each boy caught an equal amount of fish or with arbitrary numbers (the first caught three kg, the second — four kg). She easily substitutes a stereotyped solution algorithm of the problem "in parts" for the analysis of the statement, and repeatedly performs fragmentary operations which have nothing to do with the statement even after the solution method has been spelled out to her.

The replacement of a true intellectual activity with impulsive attempts, against the background of her inability to go over to the consistent analysis of a program of operations determined by the statement, is a graphic example of her disintegrated intellectual processes. It is even clearer as we deal with a highly educated patient (an engineer by profession) with, it would seem, perfectly intact mental processes.

The data described should by no means be regarded as accidental. We obtained similar results in a repeated test on this problem conducted ten days after surgery when the patient had physically completely recovered. For example, the patient is once again given the same problem. She repeats its statement correctly and again says that she does not know how to solve it. Its statement is once more written down in the form of three consecutive equations.

The patient looks at these equations and writes down: "1 part + 2 parts + 3 parts = 6 parts = 11 kilograms." *No, these are not parts but ordinal numbers, the 1st, the 2nd and the 3rd boys!* (The patient remains indifferent and does not resume her problem solving.) *Tell me what you are doing!* "One part . . . and two parts . . . these are virtually not parts but seven kilograms . . . one . . . and two . . . these are not parts . . ."

The patient once again repeatedly reproduces the stereotype of a problem in parts.

Her further attempts to solve the problem do not go beyond impulsive guessing. The patient failed to initiate a planned program of problem solving.

The patient encountered similar difficulty in working out conflicting problems. She was so much affected by the direct perception of the statement elements that she fell back on an impulsive solution algorithm which contradicted her own solution plan, even after she had actually performed the desired operations.

The patient gets the problem: *A pencil which is 16 cm long casts a shadow which is 48 cm longer than the pencil. How many times longer is the shadow than the pencil?*

While repeating the problem the patient simplifies its statement and the main question of the problem gets lost. "The casted shadow is 48 cm long... we should learn the length of the shadow and the length of the pencil. 48 + 16 = 64... The length of the shadow... that's all! (The statement is repeated again.) Aha... how many times longer is the shadow than the pencil . . . so, the shadow which is 16 cm long... is casted... on the length of 48 cm . . . how many times longer is the length of the shadow than the pencil?"

The question of the problem is replaced with the question on the length of the shadow. Later there arises the persevering question on the length of the pencil already given in the statement.

In the course of repetition the statement disintegrates.

How long is the pencil? "16 + 48... no, the pencil is much longer... wait a minute, I'll calculate."

An impulsive repetition of the performed operation.

The initial data is ignored.

The patient is shown the table:
the pencil — 16 cm long;
the shadow — 48 cm longer.

She repeats the statement correctly. *Solve the problem.* "48 + 16 = 64." *What does it stand for?* "I think that'll be 64 cm... this is the shadow... it is how many times longer the shadow is than the pencil..." *You learned how many cm and you should have learned how many times...* "Oh... how many times... then... the pencil is 16 cm long... just a minute . . . it should be several times shorter . . . Aha! Three times! 46 ÷ 16 = 3."

The obtained values are not used as the initial data for the subsequent operation and she once again tries to solve the problem by impulsive operations.

This extract reveals the same tendencies: the patient is unable to analyze the statement of the problem, to develop its solution algorithm, and then to consistently adhere to it. The hierarchy of operations involved in the general program is replaced with impulsively arising fragments which never form part of a system of planned operations. Similar results were also repeatedly obtained in subsequent experiments. Very gradually the above syndrome began to disappear and six weeks after the operation the impulsive operations began to diminish. By the time of her discharge from the hospital, the patient had relatively recovered her ability to perform structurally complex intellectual operations.

We have described the disorders in intellectual activity in the case of a nearly symptomless arachnoendothelioma located in the left frontal-temporal area. Similar manifestations could be observed when the analogous tumor was located in the superior convex sections of the left frontal lobe.

Patient D, 51-years-old, case history No. 37,604, an administrative officer, was admitted to the Institute of Neurosurgery complaining of severe headaches. Three years earlier, he had noticed the protrusion of the bone in the right posterior frontal part of the cranium; the osseous growth gradually grew larger and by the end of the third year the exostosis had reached 4 by 6 cm in size. When the tumor began to grow the patient began to suffer from headaches.

The assessment conducted in the Institute of Neurosurgery noted papilledema in the fundi of both eyes and pyramidal incompetence to the right. We observed no sensory disorders or any other neurological symptoms. The blood pressure reached 239 mm of the water column, and the blood contained 0.66% protein against a background of a normal cytosis.

The EEG showed a distinct focus of pathological activity (slow waves) in the left frontal-temporal area.

The patient's neuropsychological assessment showed no real abnormalities. At the same time, he underwent manifest emotional changes which revealed themselves in flabbiness. His gnosis, praxis, and speech were perfectly intact. He did not exhibit any gross impairments of intellectual functioning. External observation did not note any manifestations of diminished activity. The main impairment observed was insufficiently stable attention. The patient himself said that he "lacked concentration." The praxis tests revealed pronounced impulsiveness which grew conspicuous whenever he was given a complicated assignment. In these cases, the patient often omitted the required recoding of the movement and replaced it with either mirror-like or persevering movements, correcting mistakes only after some time. The patient underwent an operation on March 31, 1964. During the operation a sarcomatous arachnoendothelioma was removed from the right os frontale. The tumor had penetrated the pia matter, the muscles and the aponeurosis in the anterior sections of the frontal lobe. The removal of the tumor revealed the considerable compression of the right frontal lobe by the mass of the tumor.

The assessment of the patient's intellectual operations revealed most graphically the impaired control of his mental processes and increased impulsiveness. These problems were particularly pronounced during an examination of his problem solving processes, keeping in mind that the patient was a cultured and well-educated person.

The patient easily solved simple arithmetic problems of the type:

$$a + b = x \text{ or } (a + b) - c = x$$

although he would occasionally make some perseveration errors, and simplify the statement of a problem. For instance, this is how he repeated the following problem: *50 pine and 35 linden boards were brought to the workshop. 59 of them were used to make benches. How many boards were left?*

"50 pine and 35 linden boards were brought to the workshop. They were used to make 50 benches. How many boards were left?"

These mistakes disappeared if the problem was repeated for a second time.

But with the increased complexity of problems, which began to require a number of intermediate operations, the clinical picture, however, changed: the patient was unable to carry through the preliminary analysis of a statement and substituted impulsive operations for a logical solution.

The patient gets the problem: *A housewife uses six liters of kerosene in five days. How much will she use in a month?*

The patient repeats the problem correctly and immediately begins to solve it: "So, she must now buy six liters five times . . . then it's 30 liters!" (He does not notice his mistake.)

He does not analyze the statement of the problem. The solution is replaced with the impulsive multiplication of the given figures.

No, this is not so. You should try harder and solve the problem correctly. "But I've already said — five times by six liters . . . that'll make 30 liters."

He neither analyzes the statement nor corrects the mistake.

The problem is split into parts and each part is analyzed aloud. The patient is given an equally easy problem: *A son is 19 years old. His father is 20 years older than the son. How old are the two of them together?*

Repeat the problem. The patient immediately begins to solve the problem instead of repeating it. " We should add — that's all . . . a son is 19 years old, his father is 20 years older than the son. So, 19 + 20 = 39 years they'll be together, or we may say that the father will be 39 years old."

The scheme of the problem is simplified. The direct summation of the figures refers both to the total age and the father's age. The patient does not notice the contradiction.

What is asked in the problem? "How many years old will the son and the father be together . . . therefore, I added. The son is 19 years old and his father is 20 years older . . . so, together they will be 39 years old!" *Is this correct?* "Yes, it's correct."

As in the previous case, the patient analyzes the statement with the investigator's help after the problem has been split into separate stages. Thereafter the problem is solved correctly. But when the patient is given another similar problem he again makes the same mistake.

The patient is given the problem: *A son is 23 years old, his father is 51 years older. How old are the two of them together?*

Repeat the problem. Instead of repeating the statement the patient at once begins to perform operations: "23 + 51 = 74. . .They are 74 years old

We once again witness the impulsive solution of the problem.

together." *Are these problems alike?* "Yes, they is perfectly alike." *Did you solve the problem correctly?* "Yes, I did."

The patient does not notice his mistake.

In this case, too, the patient finds the mistake only after the problem has been split and each of its questions analyzed.

Similar manifestations could be also observed in other experiments: logical problem solving was replaced with impulsive operations irrelevant to a careful analysis of the statement. We repeatedly noted such mistakes whenever the statement of a problem was complicated. It is worthy to note that the patient never compared the obtained answer with the statement and generally did not notice his mistakes.

These facts grew particularly conspicuous in working out so-called standard problems.

The patient gets the problem: *There were 18 books on two bookshelves; there were 2 more books on one of them than on the other. How many books were there on each shelf?*

The patient repeats the problem correctly and immediately begins to solve it: " It's clear, 18 ÷ 2 = 9 . . . 11 and 9!" *How much is that altogether?* "Twenty." *And how many books were there according to the statement?* "Eighteen." *Did you solve the problem correctly?* "Yes, I did."

The comparison of the obtained result with the original statement does not lead to the realization of the mistake.

The patient is given a scheme which explains the statement. The patient looks at it and then says, " In this case it's 18 + 2 = 20."

All his further attempts to find the correct solution fail. The investigator spells out the solution algorithm to him:

1. you should first discard the surplus;
2. then you should halve; and only afterwards,
3. you should add this surplus to the number of books on one of the shelves.

To check whether the patient has mastered this technique he is given a similar problem: *There are 40 apples in two baskets. There are six more apples in one of them than in the other. How many apples are there in each basket?*

The patient repeats the problem correctly, but formulates its question in the following way: "How many apples were there in both baskets?" The investigator draws his attention to the incorrect repetition of the question, and with his help the patient repeats the statement correctly.

The question is replaced with the echolalia of the statement.

Solve the problem. "40 ÷ 2 = 20... there are 26 on one and 14 on the other." *Did you solve the problem correctly?* "Yes, I did . . . we shall take six apples from one basket and put them into the other... we shall halve 26 + 14 = 40."

The patient repeatedly solves the problem in keeping with the established stereotype. He neither compares the obtained result with the statement nor notices his mistake.

For control purposes the patient is given a seemingly similar but structurally different problem in parts: *There were 21 books on two shelves; there were twice as many books on one of them. How many books were there on each shelf?*

The patient repeats the statement correctly but omits the question. *And what should you learn?* "How many bookshelves were there in the room... no, how many books were there on two shelves."

The question is replaced with either some irrelevant question undetermined by the statement or with some information given in the statement.

The statement is written down. The patient is asked to solve the problem, consulting the statement. "On two shelves... so, there were 42 books if there are twice as many . . . no . . . on two bookshelves . . . there is something wrong here . . . (a pause) no... there seems to be 42 books."

The solution of the problem is replaced with the operation 21 × 2, triggered off by the fragment "twice as many."

The patient refuses to go on with the solution.

These illustrations show graphically enough the fragmentary nature of his impulsive attempts to solve the problem. We noted similar facts when he tried to work out problems in which it is even more necessary to compare the constituent data.

The patient gets the problem: *Three boys caught 11 kg of fish; the first and the second caught seven kg; the second and the third — six kg. How many kg of fish did each boy catch?*

The patient attempts to repeat the problem: "All in all there are four boys. They caught 11 kg. Two boys caught seven and two others — six kg." *What should we learn?* "I forgot."

The given equations are impulsively summed up, leading to the alteration of the statement.

The problem is presented again and the patient successfully repeats it.

Explain your solution plan. "So... the first and the second caught seven kg . . . The second and the third — six kg. Thus, they altogether caught 13 kg... no, they caught 14 kg all in all. . . and who accounts for one kg... I can't make out. . . ."

Despite the correct repetition of the statement, the patient repeatedly adds the terms of the two equations.

The problem is presented to him in writing. "So, what happens... three boys — 11 kg... we shall put it aside and remember... and seven kg and six that will make 13... the problem is child's play and I can't make it out... (Reads the problem.) so, the first and the second = seven kg. I would divide seven kg by two boys but that means that someone more... Therefore, the first — four kg, the second — three kg and here we shall halve six and get three kg each."

The solution of the problem is replaced with impulsive numerical operations undetermined by the statement of the problem.

An arbitrary operation undetermined by the statement.

The solution algorithm is prompted to him: *Look at the first and the second equations. What is absent in the second equation? You see, the first equation is the sum total of the three boys' catch. Who is absent in the second equation?...* "The third boy... then we must add seven and 11.... "

The patient does not take the cue and repeatedly performs impulsive numeric operations instead of solving the problem.

The described behavior of the patient appears very much like the previous patient. As in her case, this patient cannot carry through a systematic analysis of a problem's statement. As in her case, consistent solution of a problem is replaced with impulsive fragments of numerical operations, which are irrelevant to the statement, and the patient ignores their disagreement. In both cases, the patients do not notice the incorrectness of the obtained results.

This patient suffers from the disintegration of a systematic program for intellectual operations into separate, impulsive operations triggered off by some individual fragments of the statement taken out of context; although the pathological state of his frontal lobes during the first assessment did not point to possible structural disorders in his mental processes.

The facts described show that distinct intellectual impairments can also be caused by injuries to the frontal lobes which show virtually no symptoms.

Patients whose disease develops against a background of general cortical irritation and disinhibition, suffer from an inability to inhibit premature fragmentary responses. These disorders reveal themselves in the impulsive answers to separate elements of a problem and the withdrawal of the preliminary analytical-synthetic work on the statement, which constitutes the orientation basis for any intellectual operation.

The patients whose disease developed against a background of languor and inactivity revealed intellectual disorders which took the form of a passive echolalia of separate, directly perceived elements of a statement as well as its simplified and inert reproduction during problem solving activities: replacement of the question with repetition of one of the terms already mentioned in a statement.

Both groups of patients feature an impairment of analytical-synthetic work on the statement of a problem, and the replacement of its genuine solution with fragmentary operations, arising irrespective of its semantic structure, which begin to dominate their syndrome.

All these facts convincingly prove that a measurable intellectual disorder of the above type is one of the most subtle symptoms of an injury to the frontal lobes, which can be observed even in the cases when all other methods for the assessment of higher cortical functions fail to demonstrate any impairment.

Chapter 8

DISORDERS OF PROBLEM SOLVING WITH SECONDARY FRONTAL SYNDROME

Clinicians are well aware of the fact that certain lesions (growths) of the cerebellum, resulting in marked hydrocephalus, may lead to a pseudo-frontal syndrome which reveals itself in the patient's general disinhibition, behavior changes, uncritical attitude to his condition, and other symptoms structurally close to the impairments manifested in lesions of the frontal lobes.

We have grounds to believe that many such cases should be regarded as secondary frontal syndromes. Thus, a pathological process created by an obstruction of the cerebral ventricles as a result of a lesion in the posterior cranial fossa can effect the state of the brain's frontal lobes, causing their pathological irritation and dysfunction. This may result in secondary changes in the function of the frontal lobes with the ensuing secondary frontal syndrome.

There arises a logical question: if a lesion located within the posterior cranial fossa and accompanied by an acute hydrocephalus really causes the secondary frontal syndrome, which symptoms should be referred to it, and which forms of intellectual disorders may be observed in these cases? Which manifestations of disordered complex intellectual operations are revealed by such patients? To which form of the dysfunctions of the frontal lobes may we refer their pathology?

An answer to these questions could help us determine some ways for clarifying the mechanisms underlying the secondary frontal syndrome.

It is common knowledge that tumors of the posterior cranial fossa and the ensuing internal hydrocephalus result primarily in severe hypertension which, however, is not accompanied by changes typical of the secondary frontal syndrome.

Our research findings have shown that in such cases patients may exhibit well-known symptoms of general hypertension (severe headaches, marked emaciation), but the structure of their intellectual processes remains intact.

These cases, which do not cause any pronounced intellectual changes, differ drastically from cases when tumors of the posterior cranial fossa accompanied by marked hydrocephalus, lead to secondary frontal syndromes.

In the latter cases a careful study of patients' specific intellectual impairments may reveal symptoms of general disinhibition and impulsiveness similar to those observed in injuries to the *basal sections of the frontal lobe*. They sometimes differ from the symptoms caused by primary lesions of this area only in some negligible characteristic.

This group of patients generally neither inhibits impulsive answers to a presented problem nor systematically analyzes its statement. Their orientation basis for an intellectual operation is severely impaired, similar to some patients with the basal frontal syndrome. Like the latter, they immediately attempt to solve a problem, even before they have listened to the statement or fully grasped it. Their impulsive attempts to solve a problem often take the form of fragmentary operations with numbers mentioned in its statement. Moreover, these operations grow completely disconnected from the statement of a problem and are undetermined by an adequate solution algorithm.

Patients with an injury to the posterior cranial fossa and the secondary frontal syndrome differ from those with the true basal frontal syndrome only in a more critical attitude to their defect. They usually see the arising difficulties and refuse to solve problems if they encounter difficulty in working them out. Another distinction is as follows: the use of special techniques designed to check their disinhibition and draw their attention to the statement of a problem, brings about a fuller recovery of their potentially preserved intellectual activity, as compared with people with massive lesions of the frontal lobes.

Here follow two typical illustrations which show in detail the intellectual disorders of patients with the secondary frontal syndrome.

Patient P, 34-years-old, case history No. 39,054, an engineer, worked as a team leader at a research institute. He was admitted to the Institute of Neurosurgery in October 1964 complaining of static and gait disorders, dizziness, and severe headaches.

The onset of his disease (a year before he was admitted to the Institute of Neurosurgery) involved attacks of occlusive headaches and static impairments. The apparent symptoms in the fundus of the fourth ventricle (incompetence of the Y, YII, IX and X-th nerves to the right), spontaneous horizontal nystagmus, severe coordination disorders of the right extremities, papilledema of the optic fundi, and an injury to the dorsum sellae suggested an acute occlusive condition caused by a growth in the posterior cranial fossa.

In between the attacks, the patient showed no evident signs of mental disorders; during the occlusive fits he suffered from general irritation, lost his bearings, and saw hallucinations.

The patient had an operation on October 30, 1964, performed by Dr. Kandel who removed a tumor from the right half of the cerebellum with a cyst (it is unknown whether it was an angioma or an angioreticuloma).

After the operation the occlusive attacks stopped. The patient, however, began to exhibit marked changes in behavior which had been masked before the operation by his grave condition as a result of occlusive hydrocephalus.

The patient was well aware of space and time but became severely disinhibited. He would break into the conversations of other people, did not listen to directions, and lectured physicians. He refused to continue the experiment whenever he suffered even a minor set-back, easily clashed with the investigator, and was absolutely unable to control himself. His mental processes pointed to severe personality changes characterized by similar pathological disinhibition.

The assessment of the patient's position and spatial praxis does not detect any significant impairments, but he often performed these tests impulsively. He simplified and reproduced the shown movements in a mirror-like fashion and corrected his mistakes only after a time lag. Similar impairments were seen on rhythmical tests: the patient often gave a large number of impulsive and excessive taps, did not inhibit them, and forgot about the assignment.

In performing tests on selective responses, the patient attempted to perform the task prior to being given the instructions and often gave impulsive echopraxic answers. Whenever he was instructed to raise his right hand in response to one beat and his left hand in response to two beats, he often raised both hands without even waiting for the signal. The strict order to listen to the instruction and not to perform erroneous movements, however, eliminated his impulsive behavior and resulted in correct responses.

Similar impulsive responses were also noted in the assessment of the patient's optic gnosis. Whenever he was asked to name complex objects or to evaluate the position of the hands on the clockface his answers were inaccurate and impulsive, but in these cases he also could correct his mistake if the investigator provided a strict order.

Similar impairments were observed in the patient's intellectual sphere. For instance, when requested to interpret the idea of a complex narrative picture, he first impulsively misinterpreted the picture proceeding from the meaning of its separate isolated fragments, but then after a strict warning, he corrected his mistake. We noted analogous facts in the tests on retelling a story and in the study of his constructive activity.

His intellectual disorders were, however, particularly evident in the study of his *problem solving*.

To study his intellectual processes we conducted a number of tests on his problem solving.

The patient easily solved simple arithmetic problems of the type:

$$a + b = x \quad \text{or} \quad a + (a + b) = x.$$

However, he encountered some difficulty as soon as he was given complicated problems which required recoding of a statement and a series of specific auxiliary operations. The moment the patient noticed that data provided in a statement was insufficient to solve a problem, he refused to work it out. The patient is, however, often incapable of inhibiting his irrelevant associations and occasionally began substituting impulsive, fragmentary operations for the process of logical reasoning.

The patient was given the problem: *A housewife uses six liters of kerosene in five days. How many liters will she use in 30 days?*

The patient impulsively repeats: "A housewife uses five liters of kerosene in 30 days. How much will she use in 30 days?"

An impulsive repetition of the statement with the repeated number 30.

The statement is repeated again. Thereafter the patient reproduces it correctly.

Tell me how you are going to solve the problem. "Nothing to solve here. . . as everything is known here. . . nothing to learn here!. . ."

He neither analyzes the statement nor tries to distinguish the unknown value.

Read the statement attentively, what is known and what is unknown? "It is unknown how much she uses in 30 days . . . it is known how much in five days. Hence, we should learn how much she uses in one day and only after this — how much in 30 days."

After the patient's attention has been drawn to the known elements of the statement, he distinguishes the question of the problem and quickly charts a solution scheme.

And now try to solve this problem by a different method. "No, it can't be solved otherwise and, generally speaking, we shouldn't even work on this. . ." *No, it can be solved differently; just try to think how.* "Then it should be 6 × 5 = 30. . . that's it! 6 × 5 = 30 and that's all! And nothing else!" *What does it mean? What does 30 stand for?* "30 days." *Then why 6 × 5?* "Because it was six liters in five days. . . yes, that's it, 6 × 5 = 30. . . You yourself said six liters in five days. If it is in five days then we should multiply by five, . . . so, for six days. . . 6 × 5 = 30! It's clear!" *No, that's wrong.* "If it is wrong, then solve it yourself!"

The analysis of the statement and the discourse are replaced with impulsive operations with the available figures which do not form part of any solution program.

Thus, the first reading of the problem gives rise to an impulsive alteration of its statement and misleads the patient to believe that there is "nothing unknown in the problem" and "there is nothing to solve here." After the patient has been requested to distinguish the unknown value and discriminate it from what is known in the statement, he easily develops the habitual logical solution algorithm. The patient is, however, unable to switch from this solution scheme to an alternative one, and all our attempts to urge him to analyze the statement on his own, and to determine a new solution algorithm end in failure. The developed solution scheme dominates, to such an extent, that the search for an alternative solution method is replaced with an impulsive manipulation of the figures given in the statement which are not even incorporated into any consistent solution algorithm.

The attempts to replace the planned process of problem solving with impulsive operations disconnected from any program become even more evident when the patient attempts to work out a problem with a hierarchically more complex solution algorithm.

The patient is given the problem: *Three boys caught 11 kg of fish. The first and the second boys caught seven kg; the second and the third — six kg. How many kg of fish did each boy catch?*

The patient repeats the problem, then tries to analyze it out loud: "So, three boys — 11 kilograms. . . this is how much they caught together. . . and the first and the second caught seven kilograms. . . am I repeating it correctly? They all together — 11 kilograms and the first and the second — seven kilograms. No, they could not have caught so much fish!.. You should sit for 11 days to catch 11 kilograms."

Irrelevant associations during the repetition of the problem.

Repeat the problem again. (The patient repeats the problem correctly.) *And now solve it.* "No, I can't solve this problem! . . . Well, okay, I'll try to work it out. . . 11 ÷ 3 . . . that'll be three point something!. . ."

The impulsive refusal to solve the problem is followed by an equally impulsive solution which is completely withdrawn from the problem's context.

And what is all this data for? You did not use it. "God knows how to solve it. . . of course, you can't divide 11 into three without a remainder. . .maybe one caught four kilograms, then 11 − 4. . . and the rest should be halved."*Does the statement say that they caught equal amounts of fish?* "Probably, they did not catch equal amounts. No, I'm not going to solve this problem!"

The patient formulates an assumption and then, also irrespective of the statement, he halves the difference.

Pay attention to the first equation. Draw up a solution plan. "Three boys. . . caught three kg. . . no, I can't make out anything here, I'm not going to solve it."

As the patient refuses to solve the problem on his own, he is given a program which splits his activity into a series of consecutive stages and controls it.[1]

Read the statement slowly. (The patient reads.) *Divide the text into meaningful parts.* (The patient acts accordingly.) *Repeat the question of the problem.* (He repeats.) *Write the statement down.* (He does.) *Formulate the solution plan.*

The patient refuses. "God knows how it can be solved. . ." The statement is split into three parts and each equation is written down separately:

He does not profit from this tip. It does not encourage his independent reasoning.

Three boys — 11 kg.
The first and the second — seven kg.
The second and the third — six kg.

Solve the problem. "I don't know how to do it! What should I consult the statement for? I can read everything in succession and that'll do!" *Compare the two first lines.* (The third line is covered.) The patient looks at both equations. "But then the third boy is not mentioned here!" The two latter equations are similarly compared. Having distinguished each component of the problem, the patient begins to solve it: "Hence, I've found the first. . . 11 − 7 = 4. . . and it's six kilograms here. . . four and six. . . so, the second caught 6 − 4. . .that is, two kilograms. . .But it was not I who solved it, it was you who did it!"

Due to the successive externalization of the statement, the patient easily solved the problem while answering the distinguished questions. He is, however, still concerned about the fact that he could solve the problem only with the investigator's aid.

The above process of problem solving is characterized by impulsiveness since the patient first refuses to work the problem out, and then attempts to give fragmentary answers instead of a true discourse. We should also note that the patient easily follows the successively distinguished solution stages and encounters no difficulty in giving proper answers. Furthermore, he is quite aware of his inability to solve the problem unaided. This ease in mastering the given program and an adequate evaluation of the obtained result, distinguishes this patient from patients with a true frontal syndrome as described above.

The patient exhibits similar defects of intellectual processes in solving standard problems.

The patient receives the problem: *There were 18 books on two bookshelves. There were twice as many books on one of them. How many books were there on each bookshelf?*

The patient repeats the problem: "There were 18 books on the shelf. . .no, I can't. . . and twice as many. . . no, I can't."

The first repetition is fragmentary in character.

After the second presentation the problem is repeated correctly.

How are you going to solve the problem? "I shall first solve it. . . no, I won't solve it. . ." *Try to work it out.* "Nine on each shelf. . . no. . . if 2 × 18. . . As there are two shelves. . .18 . . .then there became nine on each. Is this correct? *No, it is not!* "Well, if it is not, then I don't know. . ."

An impulsive refusal to solve the problem.

Impulsively making attempts to halve. The statement of the problem is not analyzed and it

Try to reason. . . "Well, on the first and the second shelves there were twice as many. . . then, $2 \times 18 = 9$. Is this correct?" *No, it's incorrect.* "Well, then I can't undertake to solve it."

does not underlie its solution algorithm.

Repeat the statement. The patient listens to the statement and repeats it. "This. . . 2×18 and there are twice as many on one shelf."

We again witness impulsive numerical operations; another refusal to solve the problem.

Try to deduce what should be done. "What should I deduce? First minus and then to multiply. . . $2 \times 18 = 9$." *But this is a problem in parts. Try to reason.* "Aha, then there are three parts. . .no, I can't solve it. . . I don't know how many books there are on the second shelf." *And how many are there on the first shelf?* "One part." *And on the second?* "I don't know, don't ask me."

Another stereotyped repetition of the developed fragmentary operations.

The prompted solution does not trigger off the desired discourse.

The patient easily solves the problem after we have worked with him on splitting it into separate questions.

We have described an intellectual disorder in a patient with a marked secondary frontal syndrome caused by a tumor of the cerebellum accompanied by an acute hydrocephalus. We have shown that these disorders, observed against a background of his general disinhibition, resemble the symptoms of an injury to the basal sections of the frontal area. In both cases, we dealt with the general background of irritation and marked weakening of inhibition processes. Also, the injury affected both the patients' general behavior and the structure of their intellectual processes. In both cases, it resulted in an inability to refrain from premature operations, to carry out respective analytical-synthetic work on the statement of a problem, and to draw up a proper solution program. In addition, both cases showed that consistent problem solving was replaced with impulsively arising attempts, triggered off by the straightforward perception of some fragment of the problem.

Probably the only feature that distinguished this patient was the ease with which he refused to solve a problem, thereby acknowledging his inability to work it out, and the ease with which he could solve it (provided the respective program). We received similar data in the second case of a tumor of the posterior cranial fossa.

Patient K, 33-years-old, case history No. 38,814, a secretary-typist, was admitted to the Institute of Neurosurgery in August 1964 with a suspected tumor of the fourth ventricle.

The clinical picture of the disease featured, in the first place, marked manifestations of cranial hypertension (severe headache attacks, papilledema with hemorrhages in the optic fundi, and transient liquorodynamic changes).

Furthermore, there were few focal symptoms which were indistinct. We could note some symptoms of pyramidal incompetence in the form of a mimic paresis of the left facial nerve, the dominance of deep reflexes in the left extremities, pain and tactile hypesthesia to the left and numbness attacks sometimes in the left, sometimes in the right extremities. We also noted hypotonia in the left leg and some symptoms of lack of coordination against a background of general hypotonia.

These symptoms could be due to either a lesion of the deep sections of the right frontal lobe, with the ensuing pathological process effecting the temporal area or a possible injury to the ventricular system. The ease with which the symptoms disappeared and their transient character (as well as the headaches) suggested the latter.

The surgical neuropsychological assessment noted the patient's general disinhibition and symptoms of euphoria. She was insufficiently critical of her state and could not concentrate. The patient was well aware of place and time, easy to contact, and quite willing to fulfill the given assignments. She had no defects either in praxis or gnosis. While tapping out given rhythms she made mistakes which she could, however, correct. She had no speech defects—she easily recited word series, named objects, and understood logico-grammatical relationships. She interpreted narrative pictures well and easily retold stories. Her counting and arithmetical operations were intact.

On September 2, 1964 she underwent surgery on the posterior cranial fossa with the resection of the inferior middle sections of the cerebellum and the subsequent opening of the inferior sections of the fourth ventricle. The operating surgeon, Dr. Bova, found a tissue infiltration there.

Her process of problem solving was similar to that described above.

She easily works out problems of the type:

$$a + b = x \quad \text{or} \quad x + b = a$$

and easily explains their solution algorithm. The patient, however, encounters serious difficulty in working out more difficult problems and makes mistakes which are typical of this group of patients.

The patient receives the problem: *A housewife uses six liters of kerosene in five days. How many liters will she use in 30 days?*

The patient repeats the problem correctly and immediately says, "It's clear, we should divide 30 ÷ 6 = 5. She uses five liters in one day."

An impulsive solution—she neither retains the question of the problem nor notices her mistake.

What is asked in the problem? "How many liters of kerosene does she use in one day." *No, it is wrong.* "Maybe, it is wrong. . ."

The problem is read again. The patient repeats it and says, "A housewife uses five liters of kerosene. . . how much does she use in 30 days?. . . Then we should multiply. . . 30 × 6 = 180. . . " *What does 180 stand for?* "She uses 180 in 30 days."

Another impulsive solution — this time she performs an operation of multiplication with the same figures.

No, this is wrong. This is a problem on parts. The statement of the problem is read once more. The patient gives correct answers to separate questions put to her.

Now solve the whole problem. "She uses six liters in five days. . . how much does she use in 30 days. . .30 × 6 = . . ." *No, that's wrong. Try to reason.* "Six liters in five days. . . this is known. . ." *And what is unknown?* "How much she uses in 30 days. . . 30 × 6 = 180." *But why 30 × 6?* "This is how much kerosene she uses. . ." *No, this is incorrect."* The first question is: how much kerosene does she use a day? 30 ÷ 6 . . . No, no, 30 ÷ 5. . . how much does she use a day. . . and then how much she will use in five days." *Did you solve the problem?* "No, I didn't."

We witness another impulsive attempt to solve the problem by the same operation.

Still another similar impulsive attempt.

The patient is offered assistance. *Repeat the problem.* (The patient repeats the problem correctly.) *Let us assume that one five-day period accounts for one part.* "How much does she use in one day. . . 6 + 1 that will make 7, no, that won't do. . . two more days are left. . . there are seven days in a week. . . 30 ÷ 6 = 5." *No, that is wrong.* "I know that it is wrong."

The rendered assistance does not restructure the solution algorithm; another meaningless impulsive operation.

To facilitate the solution, the patient is offered another problem: *A housewife uses ten liters of kerosene in five days. How much does she use in 30 days?*

The patient begins to solve it at once: "10 ÷ 5 = 2. . . and then. . ." *What does two stand for?* "This is in one day. . ." *What is asked in the problem?* "How much does she use in 30 days. . . to multiply one day by 30 days. . . so, 2 × 30 = 60. . .60 liters of kerosene in 30 days."

The problem with the simplified statement is solved correctly.

After the simplified problem has been successfully solved, we check if the patient can transfer this technique to more difficult problems. The patient is asked to repeat the former problem. She repeats it correctly.

Now solve the problem. "First, how many liters does she use in one day. . . the 6 ÷ 5 = x. Then we should multiply x by 30 and we shall learn how much she uses in 30 days."

The patient transfers the solution algorithm to the problem with a more complicated statement.

We also check if the patient is able to solve the problem otherwise. *And now let us assume that five days stands for one part. Calculate, how much does she use in one month?* "1 + 1 = 2. . . 3 × 2." *How many such five-day periods are there in a month?* "30 ÷ 5 = 6." *What does six stand for?* "Six parts." *Then how much does she use in a month?* "6 × 6 = 36 liters."

Another impulsive attempt to solve the problem. The patient successfully applies the prompted solution technique. The problem is solved correctly.

The review of the above material proves that the patient cannot solve a complex problem since its solution necessitates a preliminary analysis of its statement and determination of a system of auxiliary techniques. Therefore, she replaces it with impulsive searching beyond the problem's context. However, when the patient is given the solution algorithm, she easily masters it and transfers it to other problems.

Therefore, we decided to check whether the mastered knowledge was sound enough and to what extent the patient was able to apply it under new circumstances.

To this end, we asked her to think of a similar problem on her own and to solve it.

Think of a similar problem. "A housewife uses seven kg of potatoes in five days. How much will she use in 30 days?"

Now solve it! "30 ÷ 7. . . No. . ." *How did we solve the problem in parts?* "We should take seven for one part. . ." *Why seven?* "Seven because there are seven days in a week."

The patient loses the meaning of the figures and reproduces, in a stereotyped way, one of her former attempts.

Repeat the problem again. (The patient repeats the statement correctly.) *What shall we take for one part?* "Seven kg. . . no, five days will stand for one part. . . it was like this: 30 ÷ 5 = 6 and there was one part there. Hence, 6 × 7 = 42 kg of potatoes are used in 30 days."

After the leading question, the patient recollects her former solution and solves the problem correctly.

Having solved this problem, the patient successfully transfers its solution algorithm to other problems of this type and solves them correctly.

Therefore, the experiment proves that the patient substitutes impulsive solutions for attempts at an independent solution. But having passed through the stage of numerous deviations, and overcoming the instability of the system she has developed herself, she masters the prompted technique and successfully transfers it to the solution of new problems.

This does not, however, mean that the patient has eliminated her intellectual impairments with the help of this training method. Our further experiments showed that moving to a different type of problem was suffient for her former impulsiveness to reappear.

The patient is given the problem: *There were 18 books on two bookshelves; there were twice as many books on one of them. How many books were there on each shelf?*

The patient repeats the problem correctly and immediately begins to solve it: "This is in parts. . . how many parts are there on one shelf. . . 18. . . parts. . . 18 ÷ 2 = 9 parts on the shelves. . . 9 + 2 = 11. . . parts. . . how many books are there on the first shelf. . .18 + 9 = 27 books. . . how many books are there on the second shelf?. . . It's a bad job!. . . On this. . . 27 . . . and there are two parts there. . .18 + 2 = 20 books. . . how many books are there on two. . . 20 + 27 = 47 . . . that's all!" *Did you solve the problem correctly?* "It seems to me I did." *What should you have remembered?* "How many books are there on two shelves?"

The patient automatically transfers the solution algorithm of the problem in parts, but then performs an impulsive operation of halving. This is then followed by an equally impulsive operation of adding books to the parts. The main statement of the problem is completely ignored.

The main question of the problem is replaced with the initial data.

The statement is presented again. The patient repeats it correctly but persists with the previous wording of the question: "How many books were there on two bookshelves?"

"I solved it incorrectly. . . it is known that there are 18 on the first shelf. . . therefore, there are twice as many on the second. Hence, there are three parts all in all. . .18 ÷ 3 = 6. . . there are 12 on one shelf and six on the other. . ." *What should you have remembered?* "How many books are there on two bookshelves?"

The patient distinguishes the number of parts and solves the problem correctly. The question, however, repeatedly reproduces the initial statement.

The problem is repeated again. The patient recites it correctly. Having repeated the statement, she begins to solve the problem once more. "I shall first halve 18. . . in this way I'll learn how many books there are on each shelf. . . as there are twice as many, we'll multiply 18 by 2. . . 18 × 2 = 36. No. . . it's wrong. . ." *Tell me what kind of problem is this.* "This is a problem in parts."

In reproducing the statement, she once again repeatedly substitutes the initial data of the problem for the question. While solving the problem she performs a number of impulsive operations disconnected from the statement.

There is one part on the first shelf. "And there are two parts on the second one. . . 18 ÷ 3 = 6." *What does six stand for?* "There are six books on the first shelf." *What should you learn then?* "6 × 2 = 12. That's all!" *What is the answer?* "There are 12 books on the second shelf and six books on the first one!"

But a single question which draws her attention to the solution method is enough to enable her to solve the problem correctly.

Having solved the problem, the patient explains its logical solution algorithm correctly.

This experiment features the alteration of the correct solution algorithm with the repeated tendency to replace the solution algorithm with impulsive fragmentary operations, which are irrelevant to the statement.

On the other hand, one prompting was enough to help the patient apply the correct solution method again, and what is most essential, to completely restore the logic of problem solving.

The facts, however, prove that this solution is not stable either. We decided to continue the experiment and give the patient another problem to determine if the impulsive fragmentary operations would occur again.

The patient gets the problem: *There were 20 books on two bookshelves. There were two more books on one of them than on the other. How many books were there on each shelf?*

The patient repeats the problem correctly and notes (in answer to the investigator's question) that it differs from the previous one and it should be solved differently. *How will you solve this problem?* "We should deduce two from 20. . . 20 − 2 = 18 books. There are 20 on one shelf and 18 books on the other. . . that's all!"	The patient begins to solve the problem correctly, but is affected by the previously applied stereotype and slides to the impulsive operation of addition.

Let us sum up the discussion of intellectual disorders in patients with a secondary frontal syndrome caused by tumors of the posterior cranial fossa accompanied by an acute occlusive hydrocephalus.

The patients of this group exhibit steady manifestations of disinhibited behavior: they are talkative, euphoric, impulsive, and easily distracted. Their entire behavior is clearly indicative of their inability to concentrate. Furthermore, these patients are insufficiently critical of their condition and suffer from an inability to evaluate their behavior.

Their intellectual disorders, manifested against this background, generally show similar symptoms. Such patients are unable to concentrate on the analysis of a statement. Instead, they immediately begin to solve the problem. As a result of their impaired orientation which forms the basis for intellectual operations, their answers are, as a rule, impulsive. They grasp some isolated fragment of the statement and begin to perform numerical operations based on that fragment. These fragments are absolutely irrelevant to the statement as a whole. They easily lose the respective denominations of figures and do not notice that their operations grow meaningless. Such patients are often satisfied with an impulsively performed fragmented operation and easily forget about the final question of a problem. They seldom suffer from the pathological inertness of established stereotypes which may be either altogether untypical of them or quite inconspicuous.

As already noted above, the symptoms exhibited by these patients resemble the clinical picture witnessed in patients with a basal frontal syndrome.

Therefore, we suggest that pathological changes in their higher nervous activity are caused by increased cerebral irritation that may be due to a dysfunction of the basal-frontal sections of the brain.

Potential intactness of such patients' intellectual processes is another feature similar to the syndrome caused by injuries to the basal sections of the frontal area. If their impulsively arising answers are inhibited and their attention is drawn to the essence of the statement, they easily grasp and master the general solution scheme and often transfer it to similar statements. Such potential intactness of their intellectual processes gives us every reason to believe that this intellectual disorder is caused primarily by their increased disinhibition, which in turn leads to severe defects in attention-concentration and impulsive premature answers.

On the other hand, we can distinguish certain features peculiar to patients with a secondary frontal syndrome. In contrast to patients with primary lesions of the frontal lobes, they are generally more aware of the problem's inaccessibility and encounter fewer difficulties in analyzing both its statement and their own operations performed in order to approach its solution. Therefore, it is easier for them to refuse to solve the problem and they do so more resolutely, showing, thereby, a more conspicuous affective reaction. As compared to patients with a true frontal syndrome, they face fewer difficulties in grasping a solution algorithm, and they usually lose the selectivity of its underlying operations. They more accurately compare the type of the given problem with the previous ones and determine both their common and distinctive features. Finally, it is much easier for them to transfer the mastered techniques to other problems. The mere inhibition of impulsive answers and then refocusing their attention to the essential elements of the statement are sometimes enough to enable them to solve a problem on their own.

We should also note that probably the most essential distinction of these patients was in the alternating character of exhibited symptoms since the experiments given them on different days yielded different results. Furthermore, in a number of cases it is enough to improve their motivation for the intellectual activity or to put them under sufficient stress to trigger off a fairly normal actualization of their intellectual processes.

We have just started to study the mechanisms underlying the secondary frontal syndrome. Only an accurate account of conditions causing this impairment of intellectual processes, and their comparison with the physiological characteristics of such patients' higher nervous processes may give us a better insight into this interesting form of higher cortical dysfunction.

---

[1] The principles underlying such a program are described in Chapter 9.

Chapter 9

EXPERIMENTS IN REHABILITATION TRAINING: PROBLEM SOLVING STRATEGIES IN PATIENTS WITH THE FRONTAL SYNDROME

In the previous chapters, we have described the impairments in the structure of intellectual processes as observed in the problem solving techniques of patients with different localized cerebral lesions. Now let us discuss the last item on our agenda—experience accumulated in compensating for the above-mentioned intellectual impairments with several different methods of rehabilitation training.

As noted above, planning for rehabilitation training should be based on different objectives and the use of different techniques since impairments of intellectual operations have different causes.

For instance, an injury to the *parietal-occipital cerebral areas* results in the impaired operations of spatial analysis and synthesis, difficulties in the understanding of the problem's grammatical structure, and in the performance of respective mathematical operations. Therefore, the main objective of rehabilitation training in this case should be to assist the patient to decode (with the help of special methods) the meaning of logico-grammatical relationships (which he fails to grasp directly) and to provide the necessary algorithms for performing the effective mathematical operations. As shown (see Chapter 2), the training programs developed for these patients permitted them to easily overcome their previously ineffective problem solving techniques.

Our experiments on the intellectual operations of patients with an injury to the *frontal lobes* gave us every reason to believe that a similar program of rehabilitation training would not produce the same results, since their perception of the logico-grammatical relationships of a statement and their ability to perform respective mathematical operations remained intact. Therefore, they did not require special training in techniques designed to analyze logico-grammatical relationships and perform mathematical operations.

The intellectual impairments of patients with lesions of the frontal lobes are entirely different in character. They are characterized by:

- An inability to carry out a logical analysis of a problem's statement.
- A loss or an impairment of the orientation basis of an intellectual operation.
- A tendency to replace the logical structure of heuristic activity with a series of impulsive operations which may either be fragmented or may reproduce the stereotype of previously performed operations, but which are always irrelevant to a statement.

We have seen that patients with lesions of the frontal lobes, as a rule, neither compare the system of performed operations with the statement of a problem nor control their performance. They do not correct their mistakes either.

Hence, we may conclude that in these cases rehabilitation training should set itself *an entirely different objective* and utilize *entirely different methods and techniques.*

Rehabilitation training of patients with lesions of the frontal lobes should not attempt *to teach methods of performing intellectual operations,* but should try *to program the patient's behavior.* Such programming should:

- Provide logical techniques for improving the patient's impaired orientation basis for intellectual operations.
- Facilitate the preliminary analysis of a problem's statement.
- Help the patient inhibit impulsively arising guesses or recurrent stereotypes of the earlier performed operations.
- Enable the patient to control the performance of intellectual operations.

These are the objectives set for ourselves to develop programs for the rehabilitation training of patients with injured frontal lobes.

Our research has, however, shown that patients with lesions of the frontal lobes do not always exhibit an identical clinical picture of intellectual impairments. Our research findings on patients with different variants of the frontal syndrome (see Chapters 3–6), demonstrate that such patients suffer from a variety of intellectual impairments.

Therefore, it is only logical that we require different techniques (or programs) for these patients to correspond to the variants of the frontal syndrome.

We discussed in Chapter 3 the principles of rehabilitation training for patients with injured parieto-occipital areas. Now let us consider the programmed education of patients with different variants of the *frontal syndrome.*

Rehabilitation Training of Patients with the Basal-Frontal Syndrome (Characterized by Prevailing Impulsiveness)

As noted earlier, the study of problem solving processes in patients with lesions of the frontal lobes enabled us to distinguish a group of patients whose intellectual processes remained structurally intact. In their cases, the correct solution of arithmetic problems was impaired only as a result of their excessive impulsiveness and apparent inability to promptly inhibit prematurely drawn conclusions. Such a disorder was typical of patients with a basal-frontal syndrome (Chapter 3).

Therefore, it was only logical that the key objective of rehabilitation training for such patients was *to draw their attention to the need to analyze the statement of a problem* and thereby, *to inhibit their impulsive inadequate answers.*

The experimental training course showed that these conditions were both necessary and sufficient to compensate for the defective problem solving techniques of these patients who could in this case successfully work out arithmetic problems. Here follows an illustration of rehabilitation training conducted with patient B whose case has been discussed in Chapter 3.

In principle, this patient was able to solve arithmetic problems. The defects of her problem solving were caused exclusively by impaired orientation of the statement of a problem. These defects often led to the substitution of impulsive fragmentary operations for the consistent and conscious solution of difficult problems. It turned out that these defects could be easily eliminated by rehabilitation training which facilitated the orientation basis for her intellectual operations with the help of certain external aids. Previous experiments with this patient showed that mere *repetition* of a problem's statement was insufficient for its correct solution, and did not facilitate the orientation basis of an intellectual operation.

To eliminate these drawbacks, the patient was asked to first *read the statement*, paying attention to its different elements and their interrelationships. Then she was asked to *repeat the problem in parts* after the investigator. We persistently drew her attention to *the final question of the problem.* Thereafter, she was asked to verbally formulate *the general solution plan* or the scheme of logical relationships inherent in the problem. Only after this analytical work was the patient permitted to begin to solve the problem.

These experiments proved that under such conditions the patient could easily solve any arithmetic problems.

Here follow the respective illustrations.

The patient is given the problem: *A father is 45 years old. He is 19 years older than his son. How old are the two of them together?*

Although the patient was instructed to first repeat and analyze the statement, she immediately answers: "45 plus 19. . . this. . . this will make. . . so, they are 64 years old. . ."

A tendency to begin the solution at once.

After she had been provided with the conditions for a consistent analysis of the problem, the patient easily solved it.

Her impulsiveness results in a solution based on a direct impression without the analysis of the statement.

Repeat the problem in parts. "A father is 45 years old (+).[1] He is 19 years older than his son (+). How old are the two of them together?(+)"

How old is the father? "Forty-five." *What else is known about him?* " He is 19 years older than this son." *What is asked in the problem?* "How old are the two of them together."

How are you going to solve the *problem?* "Well, it is clear you know, I'm so muddle-headed now, 45 − 19 = 24 years as the father is 45 and we don't know how old the son is and after this 45 + 24 = 69. They are 69 years old together."

The following day the patient was given this problem again and she failed to solve it. But when the use of the above technique provided the conditions for concentrating her attention, we thereby laid the foundation for the conscious analysis of the statement. The regulation of the patient's behavior yielded the desired result.

We observed similar results when she was asked to solve the following problem: *One boy bought one ABC book and one pen and paid 37 cents for them. Another boy bought one ABC book and two pens and paid 49 cents for them. How much do an ABC book and a pen cost separately?*

The patient could not solve this problem without assistance. Her problem solving once again revealed a tendency to operate impulsively without any preliminary analysis of the statement. Therefore, she solved the problem incorrectly, although made no mistakes when repeating the statement.

In this case, too, splitting the statement yielded favorable results.

"Let us try to find how much more the second boy paid. 49 − 37 = 12. This is how much one pen costs. Now 12 ÷ 2 = 6. And now we'll learn how much one ABC book costs. 42 − 12. . . if each pen costs six. . . then two times—12. . . 37."

The essential question of the problem is distinguished correctly. She is affected by the fragment "two pens" and performs an operation of division which is meaningless here.

After the abortive attempt to solve the problem on her own, the patient is asked to work according to the plan: to repeat the problem in parts. She fulfills the task well. Then she is asked to distinguish and underline the essential relationships between the given entities.

What is essential here? "But I've already said, 49 − 12. . . 12 ÷ 2." *Calm down and listen to me. What is the difference between the two purchases?* "Well, certainly, a pen costs 12 cents — that is essential. And I thought two pens. . . why? (laughs) Then. . . two pens — 24 cents and an ABC book. . . 49 − 24 = 37, an ABC book costs 37 cents. And that is all."

She is influenced by the perseveration. The cue is taken up.

The above records show that rehabilitation training aimed at the consistent analysis of a statement and regulation of the patient's behavior during the course of problem solving yielded good results. We should, however, note that in its verbal form such training, effective as it was, brought about only temporary progress. Whenever the patient was given a new problem, she impulsively began to work it out without analyzing its statement and the therapist often had to repeat the same instructions over and over again before she could accurately solve it.

Therefore, like in the experiments of the patients with parietal-occipital lesions, we also attempted to develop a *fixed program of operations* that would be adequate to the patient's intellectual impairment. To this end, we used *written programs of successive operations* to be performed by the patient.

The program was as follows:

1. Read the problem *attentively* and *slowly*.
2. *Divide* the text into *semantic parts* and *underline* them.
3. Repeat *the question* of the problem.
4. Write down *the statement* of the problem or its *scheme*.
5. Recite your *solution plan*, consulting the statement.
6. Solve the problem. *Compare your solution with the statement.*
7. Give the *answer* of the problem.
8. *Read the program; be attentive, don't be in a hurry.*
   Say everything aloud.

Training the patient in problem solving with the support of this program gave a relatively stable result. After the patient had mastered the recommended method of operations, she could correctly orient herself to the statement of a problem and consistently solve it using the program of operations on her own. We had, however, each time to motivate her and to remind her of the importance of adhering to the recommended sequence of operations. While the patient could usually perform a verbal program, the effectiveness of the problem solving process required an external written program which she repeatedly used to control her intellectual activity. We should also note that as soon as the written program was removed, she began to rely on impulsive operations. The patient could orient herself to the problem statement, analyze it, and control her activity only with the use of external support or written directions. Here follow some illustrations testifying to the above influence of the program on her problem solving operations.

The patient is given the problem: *A son worked 23 workday units; his mother earned nine workday units more than the son; his father worked 19 workday units less than the son and the mother taken together. How many workday units did the father work?*

The patient, as usual, immediately begins to solve the problem without repeating and analyzing its statement.	An earlier observed tendency towards impulsive activity.

"The mother worked 31 workday units. . ." *Do you remember how you should work?* "Calmly and with the program. Give me your sheet, please. It is true, I can't do without it. So, I think, 23 + 9. . ." *Just a moment. Read the first item of the program.* (She reads out: "Read the problem attentively and slowly.") "Precisely what I'm doing." (She reads the problem. The investigator has to regulate her rate of reading and then draw her attention to the data mentioned in the statement. She reads the second item.)

The patient realizes that the program helps her.

"So, I divided. A son worked 23 workday units. This is the first part. His mother worked nine workday units more — this is the second part. The father worked 19 workday units less, than the son and the mother taken together. . ." (She repeats the question of the problem correctly.) "I need not write the statement down. I already know how to solve it. We'll first find the mother's workday units and then the father's ones. No, first the mother's then the mother's and the son's taken together and then the father's workday units. Twenty-three workday units + nine workday units = 32 workday units — this is the mother's and now 32 workday units minus 19 workday units." *Consult the statement.* (She does.) "Oh, gosh. 32 workday units + 23 workday units = 55 workday units; 55 workday units − 19 workday units = 35. . . 34 workday units."

A meaningful analysis of the problem results in the correct understanding of the problem's logical structure and the development of a solution plan.

Repeat how you should work. " I should first read the problem. (thinks) Then. . . then. . .then I should solve. . .that is all." *Read the program. Memorize the sequence of operations in the course of problem solving. Next time you will work unaided.*

Having read the program, the patient repeated it correctly and consistently solved the problem.

The record shows that the support of the program has drastically changed the character of her problem solving. The program has mediated the patient's activity and provided continuous external support. Therefore, as soon as her activity has been programmed externally, the patient's work on the problems is normalized.

The next problem corroborated these findings:

The patient receives the problem: *Ann and Kate were playing baseball. They got seven points for each hit. Ann got 21 points and Kate — 35 points. How many more hits did Kate have?*

The patient repeats the problem: "Ann and Kate were playing baseball. They got seven points for each hit. One of the girls had 21 points and the other — 35 points. How many times more hits did one of the girls have?"

The attempts to repeat the statement of the problem on her own end in failure: the patient repeats the statement correctly twice, but fails to recite the question of the problem, impulsively substituting a different question for it. She does not notice her mistake either.

Did you repeat the problem correctly? "Yes, I did." The investigator draws her attention to the mistake and reads the problem once again. Thereafter, the patient repeats the problem correctly.

Solve the problem. "One girl had... three . . . hits and the other — five. That's all." *And what then?* "Then. . . I'm bad at mathematics. . . Should I divide?. . . Or should I subtract? (laughs) No, I'll subtract two from five. . . no, three from five. . . one girl had two times more points than the other." *Is it really two times more?* "Then by two times."

Her arithmetical operation is disconnected from the statement. We witness perseverations in the assessment of the denominations.

The patient is given the program. She quickly reads it without attempting to perform any operations. The investigator covers all the items of the program with a sheet of paper except for the one she needs at that moment. This technique turns out to be quite effective. As soon as the patient has read the respective item, she performs the respective operation. She reads the statement of the problem slowly. Then she correctly splits it into the semantic parts and writes the statement down in the following form:

1 (one) hit — 7 points.
Ann — 21 points.
Kate — 35 points.
Ann's hits — ?
Kate's hits — ?
How many more hits did Kate have?

While writing down the statement, the patient has already developed the general solution scheme as reflected to some extent in her notation of the statement.

The patient reads Item Five of the program which demands that the solution scheme be drawn up, relying on the information of the statement, and says, "This is difficult. . . the plan is, certainly, difficult. I'm such a muddlehead. But I think it could be this way: we should learn Ann's and Kate's hits, then their points and then the difference."

She reads Item Seven. *Solve the problem. Check the solution.* "Yes, it's true, I should check, I'm so muddle-headed. . ."

Recurrent stereotypes in the evaluation of the denominations.

Solve the problem. "21 ÷ 7, that'll be three, but hits or points?" *Consult the problem.* "Yes, certainly, hits. Then we shall divide 35 into seven and get also five hits. 5 − 2 = 3 hits. What is this? Yes, what is this? This is Kate's. . . no." *Consult your notation and the statement.* "Kate has more hits than Ann." *Did you solve the problem correctly?* "Yes, I did."

The investigator's questions help her understand the performed operation.

Patient B was trained in problem solving with the help of the program for four weeks. During this period, we not only got a better insight into the structure of the patient's intellectual disorder, but we also formed a basis for the compensation of her impairment. It turned out that the patient was capable of learning. It took her relatively little time to master the program and memorize its contents—with its help she always solved problems correctly. But her progress was only possible with the external support of the "materialized" (written) program and the continuous regulation of her work by the investigator. At this stage, all the attempts to transfer to self-regulation, with the help of the patient's speech, ended in failure. Here follows a respective illustration.

The patient is given the problem: *There were 150 crystal vases in a shop. Some of them were sold. After this, 65 vases were left. How many vases were sold?*

The patient immediately begins to solve the problem without even attempting to analyze its statement. "So, there were 150 and 65. . . that 65 were sold. . . it seems to me. . . 150 − 65—this is what was left."

Lack of orientation in the statement of the problem leads to the wrong wording of the question.

Just a moment. Don't be in a hurry. Remember the program for your work. What should you do first? "Read the problem attentively. (laughs) And I took to my heels at once. But I remember it without reading. There were 150 crystal vases in a shop. Sixty-five were sold. How many vases were left? So, 150 − 65 and that's all." *Read the problem.* (The patient reads the problem.) *What should you do now?* (a pause) "Divide the problem into semantic parts as far as I remember." *Act accordingly.* (The patient correctly fulfilled the task, but then once again tried to solve the problem.) *Just a minute, what's then?* "I should repeat the question of the problem. (Consults the text.) So, yes, how many

The incorrect wording of the statement persists.

The patient realizes that she needs a written program.

vases were sold? Or is it wrong? Give me the written program, please. (The patient consults the program, reads and executes each item.) Yes, certainly, 150 − 56 = 85. Eighty-five vases were sold."

The problem is solved correctly with the support of the written program.

The record shows that the patient remembered not only the content of the program, but also the sequencing of its items. But her own speech, however, was unable to regulate her intellectual activity.

We have observed similar findings during the process of problem solving in a number of different cases when the main impairment of the patient lay in his/her inability to perform a logical analysis of a statement and to inhibit impulsively arising fragmentary attempts.

Patient D, for instance, (see Chapter 7) featured potentially intact intellectual operations against a background of an indistinct frontal syndrome. His main impairments were distractability and apparent impulsiveness of his behavior, which led to marked intellectual impairment. Like the previous patient, he could solve simple problems well. But he often substituted impulsive uncontrolled operations for the consistent solution of complex problems, which required a preliminary orientation of a statement and the choice of a solution algorithm from numerous options. Like in the previous case, his problem solving was actualized without any preliminary analysis of a statement and replaced with ineffective impulsive operations. He generally neither noticed nor corrected his mistakes. But the investigator's unobtrusive guidance was often enough to eliminate these impairments and bring about the correct solution of a problem, although it did produce a stable basis for the compensation of the impairment. The above training program turned out to be effective in his case, too.

Here follow some illustrations.

The patient is given a problem (which he has already repeatedly worked out): *A housewife uses six liters of kerosene in five days. How much kerosene will she use in 30 days?*

Although the directions require the preliminary repetition of a problem, the patient immediately begins to solve it without making any attempts to analyze its statement.

"So, she must five times buy by six liters, therefore, 30 liters." *How did you learn this?* "Six liters in five days, 6 × 5 = 30 liters in 30 days."

Listen to the problem once more. (The problem is read for the second time.) "Everything is clear now. So, 5 ÷ 6. . . no, 5 × 6 liters." The patient talks a lot and very quickly; he is impulsive not only in his intellectual operations, but also in speech and movements. His movements are fussy.

The patient is given the program. He reads it and performs the respective operations. He correctly splits the problem into semantic parts; he looks up and begins to solve the problem.

"So. I read the problem. Aha, I should learn how much she will use in 30 days. Therefore, 5 ÷ 3, that is 5 ÷ 6. . . no, no. . ." *Work more attentively. Read Item Five again.* "So, (consults the statement of the problem) 30 ÷ 5 = 6 —

that is, she will buy six times by. . . (consults the statement) by six liters. So, $6 \times 6 = 36$ liters of kerosene she is going to use in 30 days."

It is evident that this patient also suffers from impulsive problem solving with the ensuing impairment or the complete loss of orientation of a statement. Therefore, he works the problem out under the direct impression of the statement by fragmented operations, which are semantically disconnected from it. Like the previous patient, he is also unaware of his mistakes.

The next record shows that even the correct verbal repetition of a problem does not lead to its correct solution. Being formal, the repetition is not followed by the analysis of the statement, and as it will be shown below, does not program the desired sequence of operations.

The patient was given the following problem: *A son is 19 years old. His father is 20 years older. How old are the two of them together?* Despite several attempts he could not solve it without the program.

As soon as he has heard the problem, the patient says, "We should add. A son is 19 years old and his father is 20 years older and together they are 19 + 20. . . they are 39 years old together." *What is asked in the problem?* "How old are the father and the son together? So, I added 19 years, that is, the son's age and 20 years as the father is 20 years older. The two of them together are 39."

An impulsive solution.

Repeat the whole problem. The patient repeats the statement correctly and once again finishes the repetition with the solution: "That's what I am saying: 19 plus 20 equals 39. They will be 39 years old together."

He does not understand his mistake.

Hence, the correct repetition of the problem does not result in its correct analytical solution in this case either. The wording of the question "How old are the two of them together" and the direct meaning of the word "together" have prompted a fragmented impulsive operation of addition.

The patient immediately solved the problem when he was given the program. "So. We should learn how old the two are together." (a pause)

Read Item Five once again. "So. . ." *Recite the solution plan. Consult the statement. Examine the statement, what is known and what is not?* "Yes, yes. It will be like this: 39—that is the father's age and the two of them. . . 39 + 19 = 59. . .no, 58 that is how old they are together."

The support of the program and the additional regulation of his activity by the investigator resulted in the correct solution of the problem.

The above records show that the program and the accompanying instructions of the investigator easily eliminated patient D's problem solving impairments. He quickly mastered the program and later could use it at the level of "loud" speech. But also in this case, additional regulation of his operations was needed to draw his attention to the set mode of action.

The aggravation of his condition before the operation affected his process of problem solving, too. At such moments the support of the program alone was insufficient for his training. He needed additional strict regulation of his work with the program. The investigator had to draw his attention to each item of the program and demand that the respective operation be performed. Thus, the investigator inhibited the patient's tendency to impulsively attempt to solve the problem relying on his direct impressions with the ensuing stereotyped solution algorithm.

As mentioned above, the prompted solution method did not help the patient solve the following problem: *Three boys caught 11 kilograms of fish. The first and the second boy caught seven kilograms; the second and the third — six kilograms. How many kg of fish did each boy catch?*

The patient did not take up the cues and persisted with the impulsive numerical operations semantically unrelated to the problem (see Chapter 7).

His behavior apparently changed after he had been given the program. He began to strictly adhere to the instructions formulated in the first three items:

1. Read the problem.
2. Divide it into semantic parts.
3. Repeat the question.

But then he started to impulsively reproduce the previous operations, although he had been reading the written items.

"So. I've read. Here it is. Divide — that is easy. Here is the first part, here is the second, and here is the third, and the question of the problem. Shall I repeat it orally?" *Please do.* (He repeats the question correctly. Consults the program.) "To write down the statement. But it is already written! A solution plan. It will be like this: six and seven kilograms. They caught 13 kilograms altogether. Now we shall divide 13 into three and get what we need." (He does not use the program. He speaks quickly as if in a hurry.)

He quickly and correctly divides the problem into its semantic parts.

He impulsively proceeds to solving the problem without the respective notation of the statement. He does not develop a general solution scheme either.

You solved the problem incorrectly. Let us begin from the very beginning. Read Item Four. (The patient is looking aside.) *Read Item Four aloud.* (a pause) *Read and act accordingly.* "To write the scheme. But this is precisely what I'm saying: to divide 13 kilograms." *Read Item Four calmly. Just read it and fall silent at once. Look at me.* (The patient obeys the

The patient once again reverts to the impulsive operations semantically unrelated to the problem.

He once again neither retains the program nor obeys it.

instruction.) "I got it. I must write." (He writes down the solution scheme correctly.)

Pronounce aloud everything you are writing. "Three boys — 11 kilograms. This is together, so, the first and the second boys — six kilograms." *Attention, what is essential here?* "The third boy is missing here." *And above?* "All the three boys are mentioned there. The second and the third — seven kilograms. It is essential that is missing here and mentioned there. I got everything."

The additional measures help inhibit his impulsive meaningless responses.

The patient correctly identifies the essential relationships in the statement but he can do this only with the investigator's guidance.

Read the program. (He reads.) "It is clear. My plan is as follows: we can learn how much fish the second boy caught (he points to the first and the second line in the notation) and here we shall learn. . . the first boy (he points to the second and the third lines in the notation). That's all." *Read the program.* "I have solved everything." *Read the program, Item Six.* "So I do, so I do. It is written here 'Read and compare.' Okay . . . 11 kilograms — 6 kilograms = 5 kilograms. This is the second boy. . ." *Consult the statement, don't be in a hurry.* "Oh gosh! What has become of me. . .I am always in a hurry. The third boy is missing here. The third boy caught five kilograms. So. And here. . . 2 + 5 kilograms = 7 kilograms. . .well, that's it. . .the second caught two kilograms. And the first. . . five kilograms — 11 kilograms. . . "

He does not realize the necessity to translate the solution plan into actual operations.

The support of the statement results in the correct operations.

Consult the statement. You need more control! "Yes, that's it (a pause), to subtract six kilograms. . . to subtract. . . to subtract what?" (The patient looks at the investigator.) *Consult your notation in the notebook. All three lines.* "Oh, yes, six kilograms – two kilograms = four kilograms the first boy caught." *Read the program.* "What for? It's all over." *Read it.* "So. Which is the final answer? Four kilograms." *What should you have remembered about the problem?* (He gives the correct answer.) *Look at all three lines.* "Well, yes. The first — four kilograms, the second — two kilograms, well, and the the third — 11 – 6 = 5 kilograms."

The patient once again loses the general solution scheme.

The investigator regulates his behavior and thereby, helps the patient recover the solution algorithm.

Impulsiveness.

Correct solution of the problem.

Throughout the experiment the patient was fussy and tearful, his movements were disinhibited, and he was continuously saying something. He began to perform operations without finishing the previous ones. The formed operations were semantically unrelated to the problem. The investigator had to somehow restrict all this excessive disinhibition, impulsiveness, and redundant verbal production. The objective was to inhibit the patient's impulsive reactions and to normalize his overall behavior. The given program per se was insufficient for this. To meet this objective, we had to resort to additional instructions. The problem could be solved successfully only if *a strict system of managing the patient's process of problem solving and his overall behavior was employed.*

Therefore, the review of the training program for patients with psychological disorders (observed against a background hyperirritability and weak inhibitory processes) indicates that what is needed to compensate for their intellectual impairments is an externally based system for managing their intellectual activity which controls the orientation basis of an intellectual operation and their behavior during the course of problem solving.

Rehabilitation Training of Patients with the Posterior-Frontal Syndrome (Characterized by Prevailing Inertness)

Now let us proceed to a review of rehabilitation training for patients with the posterior-frontal syndrome, described above. In this case intellectual impairments are characterized by pathological changes in *the dynamics of intellectual operations* as a result of *inactivity and inert stereotypes.*

The behavior of such patients during the course of problem solving is drastically different from patients with the basal-frontal syndrome. For example, they never initiated an activity; they had to be urged or encouraged to perform the required operations. Having performed one operation, they generally "froze" and remained inactive. An external stimulus was needed to help them go from this operation to the next one. Otherwise, their overall intellectual activity, was ineffective in the absence of continuous external *programming*. An impairment in the active forms of intellectual operations resulted in the substitution of passive inert stereotypes to process information given in a problem and distinguish essential relationships.

With this in mind we tried to develop training methods which would allow patients to compensate for their chief intellectual impairments.

Our previous attempts at rehabilitation training, partially described above, showed that training this group of patients should be designed, on the one hand, to eliminate their inactivity, marked lack of spontaneity, and initiative, and on the other hand, to eliminate their inert stereotypes and impairments in the dynamics of their intellectual processes. Therefore, the objectives and content of the programs used for training patients with the injured posterior-frontal systems differed from those utilized in rehabilitation training of patients suffering most from the impulsive attempts to solve a problem without first analyzing its solution plan.

For instance, the study of patient B (see Chapter 4) who had a tumor removed from the posterior-frontal area showed that in this case it was not enough to provide conditions for the preliminary orientation of a problem. A repeated verbal (and also later written) *encouragement to a successive series of desired operations* was needed to trigger the desired cycle of consecutive intellectual operations.

Such comprehensive external programming of his intellectual activity enabled even this patient to overcome the defects of his problem solving. First he was trained *verbally.* The patient wrote down only those investigator's directions that he was not expected to perform at once.

The patient is given the problem: *A worker paid four roubles for fur gloves, he paid five times more for a hat and the total sum of the gloves and the hat for a jacket. How much did he pay for the whole purchase?*

The patient does not begin to solve the problem on his own.

Repeat the problem. Repeat. . . "A worker paid for fur. . . for. . . the gloves. . . 42 roubles." (A pause, he does not make any attempts to go on with the repetition.) *And what then?* (a pause) *Go on further!* "Further. . . for a jacket. . . five times more." (Another long passive pause.) *Did you repeat everything?* "No, I did not." *Then why are you silent? Have you forgotten the rest?* "No, I have not." *Go further.* "Further. . . for a hat twice. . . as much. . . no, rather as much as for the hat and the gloves taken together. No, there is something wrong."

Echolalia, perseverations, and expressed inactivity.

Encouragement does help.

An assimilation of the statement elements.

The realization of his mistake.

After the second presentation, the patient recites the problem correctly (although he encounters some verbal difficulties: perseverations, echolalia) but he does not begin to solve it.

Do you know how to solve the problem? "Yes, I do." *Then why don't you solve it?* "I don't know. . . for no particular reason." *Listen to my command! Formulate the solution plan.* "It seems to be. . . in three operations. . ." (a pause) *What can you learn first? Think, consult the statement.* "I shall first learn how much he paid for the hat." (He keeps silent.) *What will you do for it?* "For it. . . I'll do. . .I'll multiply the gloves five times. . . that'll make 20." (a pause) *Why don't you solve further? Don't you know what's to be done?* "I do know. Command." *What did you learn? Repeat, please.* (The patient repeats correctly.) "I'll learn how much he paid for the jacket." (a pause) *Which operation should you perform?* "I'll add . . .these together. . ."

The solution algorithm is perfectly intact, but the process lacks activeness and dynamism. The external programming of the operations (stimulation) leads to the correct solution of the problem.

This demonstrates that the patient's ability to solve the problem is potentially intact. He is aware of the general solution scheme and the specific plan for its implementation. The impairments in his level of activity provide an obstacle for him to complete the task. As shown above, a system of successive instructions which encourage the patient to perform the operations is very effective for solving such relatively simple problems. In the cases of more difficult problems or of a more severe syndrome of inactivity, an external stimulus is needed to promote the recovery of the desired dynamics of the intellectual process by encouraging the analysis of a problem's statement, and to determine its essential relationships.

For instance, the patient was given the following problem: *A son is five years old. In 15 years his father will be three times older than the son. How old is the father now?*

It turned out that the patient could solve it only after the problem had been split into a series of successive steps which urged him to perform the necessary operations designed to analyze the statement:

- Read the problem.
- Point to what is known and underline it.
- Identify the unknown and point to it.
- Repeat the question of the problem, etc.

On the other hand, the problem could be solved if the conditions which secure the proper dynamics of the problem solving process were provided.

In the cases of a severe pathology of active processes, when the patient made absolutely no attempt to perform an intellectual operation on his own, we resorted to the following specific forms of encouragement: the *investigator began* each subsequent question of the problem and the patient had to *complete it* relying on the general solution plan. The beginning of the question by no means served to prompt the idea of the intellectual operation, it simply involved the patient in verbal activity. This method of "completing a phrase" is well-known both in the theory and practice of rehabilitation training. It is used whenever it is necessary to restore the *dynamics* of speech which may, of course, be impaired in dynamic aphasia. In this particular case of training patients with a syndrome of inactivity in problem solving, this method also turned out to be effective for restoring the dynamics of the intellectual process.

As shown above, the patient failed to solve the problem on his own: *A son is five years old. In 15 years his father will be three times older than the son. How old is the father now?*

The programmed orientation of the problem and continuous encouragement of action resulted in its correct solution.

Read the problem. (The patient reads the problem.) *Repeat it.* (The patient recites the problem correctly.) *Underline what is known in the problem.* "Well. . . the boy is five years old. . ." *When?* "Now

. . .and the father. . . will be three times. . . older. . ." *When?* "In 15 years." *Is this known?* "Yes, it is." *And what is unknown?* "How old is the father." *When?* "Now." *And when is it known?* "In. . .no, it is not known precisely." *Solve the problem.* (a pause) *You've retold everything correctly . . .* "So, I'll learn . . . this. . ." (verbal difficulties, undecipherable words) *The first question. How old...* "Will the son be in. . . that's it. . ." *What should you do?* "15 + 5 = 20 years." *Go further.* "What further. . ." *Consult the statement. How old will be . . .* "The father. . .in 15 years..." *What will you do?* "15 + 15. . .30 years. . ." *Consult the statement. How old will the son be in 15 years?* "Twenty. . ." *And how old will. . .* "the father 20 × 3 = 60 years. . ." *Did you solve the problem?* "Yes, I did." *What is asked in the problem?* (The patient repeats the question correctly.) *How old is he now. . .* "the father. . . 60 years. Is he. . ." *When will he be 60?* "He is 45 years old now. . .I wanted to say, 60 − 15 years."

The patient is involved in speech.

An inert stereotype.

Perseverations.

Externalization of the internal relations of the problem to the realization of the performed operations.

The perseveration error. The correct solution of the problem.

It is evident that the patient is able to solve the problem in principle. But he can solve it only if he is provided compensation of the main impairment in his active intellectual activity. That is, when the process of problem solving is divided between two people—the patient and the investigator—the investigator takes the lead.

After we determined the main direction and specific rehabilitation training methods for patients with such disorders, we were able to develop a written program for externalizing the patient's operations which regulated the process of his problem solving.

The external programming of intellectual activity for this group of patients laid the basis not only for a short-term compensatory effect, but also for a long-term training effect.

The program had to meet the following requirements:

- To strengthen the orientation basis of an operation.
- To provide the conditions for a systemic series of operations designed to solve the problem.
- To secure the patient's self-control over his actions.

Therefore, the program included a series of directions which had to be executed by the patient. In order to draw the patient's attention to the most important items, they were typed in red. Here follows a model program.

1. *Read* the problem.
2. *Divide* it into semantic parts and demarcate them with a line.

3. *Write* these parts *down* one under another.
4. *Underline and repeat what is asked in the problem.*
5. Solve the problem.
6. Can you *answer the question of the problem?* If not, then...
7. *Look at the statement of the problem attentively* and find what is unknown in it.
8. How can the *unknown* factors be found? Write down the *first question and perform* the required *operation.*
9. *Compare* it with the statement.
10. Determine if you have answered the question of the problem. If not, then...
11. Write down *the second question* of the problem and *perform* the respective *operation*.
12. *Compare* it with the statement.
13. Determine if you have answered the question of the problem. If not, then...
14. Write down *the third question* of the problem and *perform* the required *operation*.
15. *Compare* it with the statement.
16. Determine if you have answered the question of the problem. If yes, then...
17. *Draw the general conclusion:* what is the answer to the problem?

This table *programs the patient's activity*. It *programs his behavior* without pointing to the specific methods for solving a problem or to the methods for performing concrete operations.

We have assumed that if this program corresponds to the structure of such patients' intellectual disorders, it will promote the elimination of their intellectual impairments and the recovery of their ability to solve arithmetic problems.

Our research findings have confirmed this hypothesis. The detailed and materialized form of programming the patient's behavior contributed to the elimination of his impairments in problem solving and helped him perform the desired series of intellectual operations.

At the first stage of rehabilitation training, characterized by the minor manifestations of inertness and inactivity, the patient could be easily trained and the program was quite effective in eliminating his intellectual impairments. Later, when his syndrome of inactivity and inert stereotypes grew much more conspicuous as a result of the continued growth of the tumor, the work with the patient began to require considerable additional monitoring on the part of the investigator, although the program continued to regulate his behavior. Here follows a respective illustration.

The patient is given a conflicting problem: *A pencil which is 16 cm long casts a shadow which is by 48 cm longer than the pencil itself. How many times longer is the shadow than the pencil?*

The patient refuses to solve it.

Solve the problem. "48. . . we'll divide. . . into 16. . ." (a pause) *Carry on.* "No,. . . 48 − 16. . ." (a pause) *What will you learn by it?* "What shall we learn. . . how much longer is the shadow than the pencil." *Go on.* "That's all."

Fragmentary operations irrelevant to the context of the problem.

Draw the schema and write down the statement. (a pause) The investigator had to begin to draw the schema himself. The patient finished the drawing. He drew it correctly:

The patient's inactivity.

16 cm

———————

the pencil

——————— ———————

the shadow 48 cm

The patient silently writes "64." *What did you learn this way?* "How much longer is the shadow than the pencil. (a pause) And that's all." *Look at your drawing. What is asked in the problem?* The patient reads the statement. "Yes, that's it. . . in 64 times." *Really?* "No,. . . by 64 times."

A correct operation which is, however, unconnected with the idea of the problem.

The inertness of the construction "by 48 cm" leads to the contamination "by 64 times."

The patient is given the program. He reads each item and acts accordingly. But he does not initiate any operations after reading Item Seven which demands that the unknown be identified.

Determine the unknown. (a pause) *Find and underline the unknown.* "We don't know. . . the length. . . of the shadow. . ." *Read the next item, Item Eight.* (The patient is reading it to himself.) *Read it aloud. Perform the operation. What should you do?* "Add." *Why?* "This 16. . . this is the projection of the pencil. Forty-eight — this is what is bigger than the pencil." *Read the next item.* The patient reads, compares his operations with the statement and on his own proceeds to read the following item. "The second question. . .48 ÷ 16. . ." *Why 48? Look at your first operation.* "Yes, 64 ÷ 16 = 4. The shadow is four times longer than the pencil." *Did you solve the problem?* "Yes, I did."

A conscious attitude to the performed operations.

Thus, the patient solved the problem with the help of the written program and the investigator's additional questions and directions.

The following record shows that an effective solution of a problem requires not only the support of the program, but also a written solution scheme.

The patient is given the problem: *Three boys caught 11 kg of fish. If we add the catch of the first and the second boys it will make seven kg, while the total catch of the second and the third boys amounts to six kg of fish. How much fish did each boy catch?*

The patient failed to solve the problem without the program and the investigator's help. Echolalias, perseverations, and inert stereotypes resulted either in inadequate intellectual operations or the misinterpretation of correct ones.

The patient begins to solve the problem: "I'll subtract seven kg from 11 kg. . . seven from 11. . . this is the third. . . the second and the third. . . six kg. . . that'll make five. That's all." *Which boy caught five kg?* "The second one." *Is it really the second boy? Consult the statement.* "The second. . . of course, the second and the third — six kg. Therefore, five kg will go to the second boy." *Go further, please.* "Further. . .11 − 9 = 2 kg — the first." *Now don't say anything at all. Read silently Items One and Two and fulfil the assignments.* (The patient correctly splits the problem into the respective semantic parts.) *Read further.* "Further — to write down. I'm writing down." *Represent the statement in the form of a table. Write the data out in the form of equations.* The patient could not fulfill this assignment. The investigator gave him a prepared scheme of the statement:

He mentally performs the required operations but does not understand what he is doing.

The echolalic assessment of the obtained result.

The patient derives the denominations of the boys from the analysis of the statement. He inertly reproduces the stereotype of the natural numerical series 3–2–1.

1. Three boys — 11 kg.
2. The first and the second — seven kg.
3. The second and the third — six kg.

Read the next item: what is unknown? The patient consults the table: "Everything is unknown." *Look at the first and the second lines. What is unknown here?* "The third boy. I got it. Seven kg — 11 kg . . . four kg. . .the third boy caught four kg."

Carry on reading. "Oh, I know. . . I should perform the next operation." (a pause) *Look at the third line. What is unknown here?* "Yes, . . . six kg of the third boy — four kg. . . two kg. The second boy caught two kg." *Go on.* "Eleven kg − six kg. . . the first boy caught five kg. . . that's all."

The record shows that the solution of the problem is facilitated not only by the system of repeated directions which programs the patient's behavior, but also by the elimination of his inertness and inactivity due to an easier identification of the essential relationships with the support of a graphic schema of the problem.

It is noteworthy that the transition from a verbal to a written solution of a problem makes it much easier to determine the desired solution algorithm. In this case the inert stereotypes, which are most pronounced in the patient's speech, exert less influence on his problem solving, and therefore, it is easier for him to adhere to the developed solution algorithm. We should specifically note that the *solution method developed by the patient is preserved fairly well* and the patient transfers the mastered principle to other similar problems and solves them correctly.

This patient also mastered the general approach to problem solving given him in the form of a program. In the subsequent experiments on rehabilitation training in problem solving, he read only some items of the program. Others he reproduced from memory.

Hence, the experimental rehabilitation training course has deepened our knowledge of this intellectual disorder. The patient has recovered an ability to solve problems by using methods which were adequate to compensate for his impairment. Among them, special note should be made of the method of *externalizing the program for the patient's operations* which was designed to regulate his behavior and eliminate his inactivity. Similar information on the *character of intellectual impairments* in patients with the syndrome of inactivity and inert stereotypes, and also on the *methods for their rehabilitation training* was obtained in experiments with other subjects from this group of patients, in particular with patient U (see Chapter 5).

We are going to analyze our experiments in training this patient since we could observe, as shown above, three different syndromes which superseded each other during the course of our observations. Our research findings demonstrated that the methods designed to compensate for his defects and his training ability varied depending on the progression of his disease.

During the first and the second stages of our research, we used some elements of rehabilitation to get a better insight into the structure of his intellectual pathology, and came to the conclusion that it was feasible in principle to train him in problem solving. But the use of the same techniques in the third period of his disease, following the bilateral lesion of the frontal lobes after the removal of a massive tumor from the medial sections of the left frontal lobe, was ineffective.

The two first periods revealed a marked *syndrome of inactivity* with a tendency towards stereotyped operations against a background of general affective changes. Our research findings show that during this time the patient could still subordinate his activities to an externally set program and retain it.

The training methods applied during these stages of his disease were similar: they were designed to encourage the patient's activity and provide external programming of his behavior. Their only difference was that during the second period the patient needed stronger encouragement to act and move to the next operation. Therefore, the system of methods applied at this time envisaged an introduction of "by-element" stimulation, discussed earlier in the review of patient B's training.

In the first period of patient U's disease, we conducted a number of experiments while training him in problem solving. Each of them testified to the possibility of compensating for his impaired intellectual operations. Furthermore, they showed that the patient was able to master a specific method for performing operations and was able later to use this method on his own. These experiments are described in Chapter 5. In this section we shall limit ourselves to one illustration.

The patient is given the problem: *A hunter sold to a shop one squirrel pelt and one marten pelt and got 50 roubles for them. Another hunter sold three squirrel pelts and one marten pelt and got 70 roubles for them. How much do one squirrel and one marten pelt cost separately?*

Repeat the problem. The patient repeats the statement and the question of the problem correctly. To draw his attention to the statement of the problem, he is asked a series of consecutive questions (which he answers correctly, too).

THE QUESTIONS	*THE PATIENT'S ANSWERS*
• *What is asked in the problem?*	+
• *What did the first hunter sell?*	+
• *How much did he get for it?*	+
• *What did the second hunter sell?*	+
• *How much did the second hunter get?*	+
• *What is asked in the problem?*	+

Solve the problem. (a long pause) *Why don't you solve it?* "I don't know." *But you don't even try to solve it. Say everything aloud.* "And what am I to say, my head is empty, nothing to catch on." Another long pause. The patient does not make any attempts to solve the problem. He is lying passively and looking away from the investigator. *Are you thinking over the solution?* "Yes, I am." *What are you thinking about?* "I don't know."

Severe inactivity in behavior.

Repeat the problem. (The patient again repeats the problem correctly.) *Write down the statement.* (The patient writes it down correctly.) *Look attentively at the data on the first and the second hunters. What do you see?* "The difference." *Answer the first question of the problem.* (A pause, the patient is diverted by an outside noise.) *You mentioned the difference: what does it lie in?* "Yes, the difference (he is writing) lies

The repetition of the problem is formal; it does not result in active searching.

The patient has a solution scheme but he needs continuous encouragement to act.

in two squirrel pelts."(a pause) *Go on*. "The difference is 70 − 50 roubles = 20 roubles." (Another pause. He once again needs encouragement.) *What does it mean? Carry on, please*. "It means that a squirrel pelt costs 20 ÷ 2 = 10 roubles. (a pause) And then we shall learn. . . how much a marten pelt costs. . . 50 roubles − 10 = 40 roubles."

And can it be solved otherwise? "Yes, it can." (a pause) *Then why don't you solve it?* "And what should I do?" *How else can you learn the cost of a marten pelt?* "70 − 30 = 40 roubles." *Why 30?* "Because this hunter had three squirrel pelts."

He understands what he has done, but shows it only in answers to the investigator's questions which stimulate him.

This illustration shows that it's not enough to simply strengthen the statement of the problem. The patient needs continuous encouragement. Only in this case can he begin to solve the problem. With each subsequent experiment the method for problem solving was secured better. Toward the end of the training course, he solved all similar problems correctly provided his behavior was externally programmed. He was still unable to solve a problem on his own unless he got additional encouragement. Therefore, we continued his training with the support of an external (written) program which was structurally identical to the program used at an earlier stage of training. Our objective at this time was to train the patient in *the general method of work* and to secure it in such a way that the patient could use this program on his own at a higher speech level.

The following experiments proved that the prolonged training could help the patient master this program and independently use it from memory (at the level of loud speech). But even in this case, however, he required continuous regulation of his performance by the investigator. Here follows an illustration.

The patient is given the problem: *A housewife uses six liters of kerosene in five days. How much kerosene will she use in 30 days?*

Repeat the problem. "A housewife uses so much kerosene in so many days." (a pause) *What is asked in the problem?* "We should learn how much kerosene she uses in one day." *Did you repeat the problem correctly?* "Yes, I did." *Try and remember, maybe you've repeated it incorrectly?* "I don't know." (The investigator reads the problem once again.) *Solve the problem.* (a pause) *Are you thinking about the solution method?* "No, I'm not thinking about anything special at all." *Do you remember how to solve the problem?* "No, I don't." *Solve it again.* (a pause) *What should you learn first?*

The patient does not know that the problem is unfinished. The statement is reproduced in a general form.

The final question of the problem is replaced with an intermediate one. The patient does not notice the mistakes made in the repetition of the problem.

An apparent inactivity in the patient's behavior.

This record shows the patient's marked inactivity, an apparent lack of dynamism and initiative. All these features are revealed already in the repetition of the problem when he substitutes an intermediate question for the final one. The patient did not begin to solve the problem despite repeated encouragement by the investigator. The latter's speech no longer regulated his activity. After abortive attempts to encourage the patient to solve the problem, he was given the above program.

The patient read all its items aloud, but nevertheless, did not initiate any activity. Then the investigator covered all the items of the program except for the first one with a sheet of paper. After the patient had performed the respective operation, the next item was uncovered, thereby, stimulating him to the next operation. This additional encouragement made it possible to draw the patient's attention to the specific item of the program, which resulted in the performance of the desired operation.

Read the program. "That's what I'm doing. What then? (He tries to uncover the next item on his own.) Aha: Divide the problem into its semantic parts and demarcate them with a line. That's it. I did divide. (He divided the problem into the semantic parts correctly. Then he wrote down the required parts.) I already know how to solve the problem." *Solve it.* (a pause) *Read Item Four of the table.* (The patient reads it and performs the respective operation.) *Let us proceed to Item Five.* The patient reads it and answers that it is impossible to find the answer, but that we should learn how much kerosene the housewife uses in one day. He writes the question and performs the respective operation. "And now the second question." (He consults the table and reads out: "Write down the second question of the problem.") "So, the second one — how much does she use in 30 days? 6 ÷ 5 × 30." (He reads further: "Did you answer the question of the problem?") "Yes, I did." *Can this problem be solved otherwise?* "Yes, it can." *Read Item Seven of the program.* The patient reads and says, "By the way, it is unknown how many five-day periods there are in a month." (He correctly solves the problem by the second method.)

The program regularizes the patient's behavior.

The patient develops a solution plan. His inactivity, however, prevents him from performing the respective operations without external stimulation.

The support of the program leads to the correct solution of the problem. Strict management of the patient's behavior provided the conditions for understanding the logical relationships in the problem and helped him to see that it could be solved by a different method.

The program resulted in the effective solution of the problem. He could not have solved it without this external support. The program stimulated and directed his activity.

The program was equally effective for solving more difficult problems.

For instance, on his own, the patient was unable to solve the problem: *A son is five years old. In 15 years his father will be three times older than the son. How old is the father now?* Therefore, he asked for the table. "Give me the table." He read it aloud and fulfilled everything. He did not stop at Item Five ("solve the problem"), but immediately proceeded to Item Six and paused.

Can you answer the question of the problem in one try? "No, probably not." *What can you learn first?* "The first question (consults the table) is how old the son will be in 15 years. 15 × 5." *Why are you multiplying?* "I don't know why, it should be this way." *How old is the son now?* "Five." *And in 15 years?* "He'll be 20 years old . . . 5 + 15 years . . ." (Read the table further.) "The second question is how old the father will be when the son is 20 years old . . . He will be 60. The father is three times older. 20 × 3 = 60." (He reads the table further. A pause.) *Did you answer the question of the problem?* "No, I did not." *Read the table.* "How old is the father now?" *How can you learn this?* "I don't know." *Consult the written statement.* (a pause) *Look at your solution. What did you learn in the first question?* "How old the son will be in 15 years." *What did you learn in the second question?* "How old the father will be when the son is 20 years old." *And when will the son be 20 years old?* "Oh, the father is 45 years old, 60 − 15 = 45. . ." *Go further.* "What further?" *Solve the problem.* "I've solved it."

The correct question does not trigger the desired operation.

After the patient's attention has been drawn to his mistake, he corrects it.

The additional questions eliminated the patient's inactivity and brought about a conscious solution of the problem.

The following records also prove that the patient is, in principle, able to solve arithmetic problems, but needs the support of the program. As will be shown below, it not only regulates his behavior, but also allows him to develop an insight into the performed operations.

The day after the patient had solved the problem: *A housewife uses six liters of kerosene in five days. How much kerosene will she use in 30 days?* he was given an analogous problem: *Some workers laid seven pipes in three days. How many pipes will they lay in 12 days?*

After the correct repetition of the statement, the patient immediately answered: "Probably 28." *Explain how you've solved the problem.* "You are being nosy." *What was the first question and the operation?* "I had no questions." *How did you solve the problem, then, in one try?*

"Yes, in one try. I just multiplied by four and that's all." *Where did you get four from?* "I simply felt like this."

The patient does not understand the solution algorithm.

We see that the problem was mentally solved correctly. The patient is, however, absolutely unaware of what he did in order to solve it. The performed operations are meaningless to him. After he had been asked to solve the problem, once again with the support of the written program, he developed a clear insight into all the performed operations.

Fragmented solution.

He does not understand the solution algorithm.

The program is mastered.

Read Item One of the program. "I already know all this. I have already read the problem, I know it and know its semantic parts . . . and what comes then. . . what is known. . . it is known that seven pipes are laid in three days. And the first question. . . this is a problem in parts." (a pause) *Where are the parts here?* "The first part — three days." (a pause)

What should you do, consult the program? (He reads: "Write the first question . . . perform the operation." *Well?* "Well, well. Three days account for one part." *And what then. . .?* "How many parts are there in 12 days. . . four parts." (a pause) *Read the program.* "Yes, I know, if I did not answer the question of the problem I should write the second question." (He is saying all this from memory.)

Correct understanding of the problem's structure. An encouragement.

Another encouragement. The patient works with the program from memory.

"Thus we can learn how many pipes the workers laid in 12 days. $4 \times 7 = 28$ pipes. That's all."

Correct solution of the problem.

This protocol is noteworthy because it reveals the importance of the program for restoring the detailed process of problem solving as realized at the conscious level. It demonstrates that the patient has mastered the program and can use it to regulate his behavior. But in this case, too, the effectiveness of the program requires continuous encouragement.

These features are particularly pronounced in the following illustration.

The patient is given the problem: *A girl found 61 mushrooms. Another girl found 14 mushrooms less than she did. The third girl found as many as the two of them together. How many mushrooms did all the girls find together?*

The patient refuses to solve the problem saying, "I don't know, it's difficult, solve it yourself." No requests or even demands to begin to solve the problem would help.

The patient's complete inactivity leads to the refusal to solve the problem.

The patient is given the program and is asked to solve the problem by sticking precisely to all its requirements. "I've read the problem. Shall I divide it according to sense?"

The patient has mastered the program and can correctly reproduce some of its items from memory. The operations are correct; the patient simply needs encouragement.

Read the program. "I know it without reading, I've divided." (The patient divides the problem into the semantic parts correctly. Then he goes on reading the program.) "Aha, I should learn what is unknown in the problem. Okay. This is unknown. How many mushrooms did the second girl find? 61 − 14 = 47. She found 47 mushrooms." (a pause) *Go further.* "What is further?..." *Read.* (He reads: "to compare") "Well, I'm comparing: everything is correct." *Go further.* "Further... How many mushrooms the third girl found. 61 + 47 = 108." (a pause) *Well and what further?* "And further..." *How many...* "mushrooms they found altogether: 61 + 47 + 108... you'd better count it yourself, it seems to total 208."

The problem is solved correctly and the patient does not encounter any special difficulty in the course of problem solving. The investigator simply had to regulate his performance with the program.

The problem is solved correctly.

Hence, it is clear that with the help of a detailed program, the patient can successfully solve the problems he could not work out prior to the training course. At the beginning of the course, the work with the program required additional regulation of the patient's activity by the investigator. Structurally, this regulation boiled down either to additional encouragement or to the additional mention of the need to use the written statement of a problem. Most often the investigator had to remind the patient of the need to control his intellectual operations.

Such was the general picture of patient U's training during the second period of his disease. Towards the end of this period, his condition exacerbated and we observed an apparent increase in his symptoms due to the continued growth of his tumor. As a consequence, the progress brought about by training became unstable and was negligible shortly before his last surgery.

Just prior to his last surgery, his neuropsychological assessment and a prolonged follow-up showed that the overall structure of his intellectual activity had changed, and that he was suffering most from a severe disorder of selectivity. Rehabilitation training had become virtually impossible for him.

Preliminary tests conducted at that time (they were described earlier: see Chapter 5) revealed that the patient had developed impairments in retaining the given program and subordinating his behavior to it. Therefore, the programs used during the earlier stages of his disease were ineffective. During this latter stage, it was no longer sufficient to remind him of the need to orient himself to the statement, encourage him to action, or to regulate his behavior. Here follows an illustration.

The patient was given the problem: *There are 18 books on two bookshelves, but they are unequally distributed. There are two more books on one of them than on the other. How many books are there on each shelf?*

THE EXPERIMENT BEFORE THE THIRD OPERATION

The patient repeats the problem correctly. *Solve the problem.* "It's in parts . . . I'm not sure, it looks like in parts . . . there are 18 on one bookshelf and two more books on the other. . . that makes 20. . . but there were 20 all in all. . . no, there is a catch here." (refusal) *Does this problem resemble the previous one?* "Both yes and no." *What is similar?* "They are about the same things — books and bookshelves." *And what is dissimilar?* "It's 'two more books' here and 'twice more' there. This is different."

He repeats the previous problem in parts.

Solve the problem. "Let's assume that there are 11 books on one bookshelf . . ." (refusal)

An attempt to solve the problem by inspection. The problem is solved.

The patient is given assistance: *Let us take two extra books away from the shelf. How many books will be left then?* "Equally . . ." *And how many will that make altogether?* "18. . . no. . .16. . .then, there are eight and ten books."

THE EXPERIMENT AFTER THE THIRD OPERATION

The patient repeats the problem correctly. *What will you learn in the first question?* "Nothing. . . there are seven on one bookshelf and 11 on the other." *How did you learn?* "I found the number."

Solution by inspection.

The patient is given an auxiliary drawing illustrating the statement (it does not affect him in any way).

Okay, you came and took two extra books. How are the books distributed on the shelves afterwards? "Equally." *Then how many books are there altogether?* "An equal amount." *And how many are there on each bookshelf?* "Eight." *Well, and then you put two books back on one of the shelves. What's the result?* "Eight and ten."

Now will you try to retell how you have solved the problem? "I don't know... have I solved it?"

The patient did not understand what he was doing.

Solve it again. "Eight and ten." *How did you get these figures?* "I don't know." *Which is correct: seven and eleven or eight and ten?* "Both."

The patient can't selectively assess whether the problem has been solved correctly.

This extract graphically shows that training (even with the complete externalization of the solution algorithm) did not result in the proper discourse, and that the patient's attempts to "solve" the problem were limited to solution by inspection.

Subsequent experiments in training the patient in problem solving ended in failure. He could grasp neither the general solution method nor the specific techniques of logical reasoning prompted to him.

Here follows an illustration.

The patient is given the problem: *A son worked 23 workday units. His mother worked nine workday units more, and the father worked 19 workday units less than the two of them taken together. How many workday units did the father work?*

Identify the first question and perform the first operation. "Which question?" *What will you learn first? Consult the problem.* "23 + 9 = 32... This is how many the mother worked." *What will you learn next?* "I don't remember, I don't know." *Consult the problem.* "That's what I'm doing. What then?" *Look, what else is unknown?* "Everything is unknown. That's the point."

Complete inactivity. The patient performs the first operation correctly.

He doesn't attempt to solve the problem, but to recollect its solution algorithm.

The patient repeats the statement and the question of the problem correctly. *Solve the problem.* (a long pause) *Consult the program.* "What shall I consult it for? I already know it." *What should you do now?* "I don't know." *Consult the program.* "Divide into the semantic parts." *Do it.* "But what shall I do?" *Act according to what you have just read.* "And what have I read?" *Read this item of the program once again.* (The patient is rereading the previous program item. A pause.) *Well, let's divide the problem into its semantic parts together.* (Together with the investigator the patient divided the problems into semantic parts correctly.) *Go on.* "What do you mean by this 'Go on?' " *Read the program.* (The

The program does not direct the patient's activity. He needs additional encouragement.

patient reads the item on distinguishing the unknown.) "Well, it's your headache."

Hereafter the patient continued to read the respective items of the program, but did not perform the required operations.

It took many attempts to help the patient solve the problem correctly, even with the help of the program. But even after all these efforts, he did not develop an insight into the generalized solution algorithm. Each time he performed the prompted operations correctly, but he did not synthesize them into a general solution plan. Therefore, he could not comply with the investigator's request to explain how he had solved the problem and said, "I neither know nor remember it."

The above protocols show that the patient could not be trained during the last period of his disease.

Hence, the results of the experiments in training patients with the posterior-frontal syndrome in problem solving demonstrate that it is practicable to offset their intellectual impairments by developing internal programs and controlling their subsequent intellectual operations, if they are given an externalized program for operations. This method formed the basis for compensating for their intellectual impairments, and in our view, was quite effective. However, the aggravation of the syndrome, as well as the development of severe inactivity and defective selectivity, made it much more difficult (and sometimes virtually impossible) to train such patients in problem solving.

Rehabilitation Training of Patients with Massive Lesions of the Frontal Lobes and Disordered Selectivity (Selective System of Connections)

We have so far been discussing the results of the experimental rehabilitation training of patients with two variants of the frontal syndrome.

We have shown that entirely different methods should be used to restore the ability to solve arithmetic problems in the case of:

1. Patients whose disorder of the orientation basis for intellectual activity revealed itself against a background of increased impulsiveness.
2. Patients suffering most from pathological inactivity and a tendency to substitute inert stereotypes for true heuristic activity.

Our research, quoted above, has shown, however, that the use of training methods enables patients of both groups to make progress and compensate, to some extent, for their impairments.

The opposite is true of our attempts at rehabilitation training of patients suffering from massive (most often bilateral) lesions of the frontal lobes, which cause a severe impairment in selectivity (selective connections). Their rehabilitation training usually proceeds with great difficulty and is sometimes altogether impractical.

The training methods described are designed to *program the patients' behavior externally* and to *stimulate the inhibited dynamics* of their intellectual processes. Our experience shows that such methods are inadequate for this group of patients. Our research (see Chapter 6) shows that these patients suffer most from the *disorder of selectivity (the selective system of connections)*. Therefore, their basic intellectual processes, which generally provide the distinguishing of logical relationships inherent in a problem and the retention of the hierarchial relations resulting from the general solution strategy, were severely disturbed. Hence, it was necessary to develop innovative approaches to the rehabilitation training of this group of patients, since we felt that they could make progress if they recovered not only the proper sequencing of the required operations, but also *the specific forms of analysis and problem solving*. In addition, it is only logical that they encountered greater difficulty in rehabilitation training, which often failed to yield the desired results.

We have come across some evidence that patients with massive lesions of the frontal lobes could not be trained in problem solving (e.g. the third period of patient U's disease). But this conclusion was most graphically corroborated by our long-standing attempts at rehabilitation training with patient B, described earlier (see Chapter 6).

We would like to supplement the above information on this case by providing a more detailed description of this patient's training program and the obtained results.

Within a few days, patient B was repeatedly asked to solve one basic problem with several modifications. We used a number of techniques designed to help him distinguish the *basic system of relations* from the information provided in the statement. The patient was taught to schematize the statement by distinguishing the relationships of its individual components. The data comprised in the schematized notation was *underlined*, and we specifically drew his attention to it. This method was supplemented by drawing schemata corresponding to the basic logical relationships. Only by following all of these preliminary operations was the patient allowed to retell the problem with the support of the schema and to distinguish its essential relationships. In the course of training, the investigator was continuously stimulating the patient and asking him leading questions. But even such assistance was often insufficient, and the investigator had to explain the interrelationships between the elements of the problem using real objects, that is, he specified the idea of the problem. If this aid also failed to meet the objective, he resorted to straightforward prompting, which it would seem, should have promoted the identification of the semantic system of relations and made the subsequent process selective in character.

The facts, however, show that even such prompting did not always guarantee the correct solution. The patient was sometimes able to repeat the proper solution procedure after the investigator, and even transferred the solution algorithm to similar problems, but he could not secure the mastered solution algorithm and repeated his mistakes during the following session. The patient was extremely impulsive in all his actions. He often began to solve a problem even before he finished listening to (or reading) its statement. His solution did not go beyond the stage of fragmentary, meaningless operations. The earlier described system of methods helped inhibit the patient's impulsive responses and focused his attention to the analysis of the statement. The use of these techniques and aids made it possible to teach the patient to independently divide a problem into its semantic components and to schematize its statement. But in this case, however, the regulation of his orientation basis was insufficient to offset his impaired problem solving techniques. The correct operations of orientation of a problem were followed by hazardous operations irrelevant to its semantic and logical systems. His severe impairment in selective connections resulted in the defective processes of abstraction and generalization, and thereby, in an inability to understand the logical relationships inherent in a problem.

Therefore, the objective of rehabilitation training, in this case, was not only *to program the patient's behavior,* but also to provide additional conditions to compensate for his impairment in selectivity.

Our attempts to compensate for this impairment ended in failure. The patient did not accept the assistance offered to him by the investigator. Even overt prompting did not help him develop an insight into the logical relationships inherent in a problem.

With this in mind, it is clear why we could, at best, train the patient in a specific method for solving each particular problem. We failed to train him in the *general mode of action* and *the general method for logical reasoning* in solving similar problems. He did not transfer the mastered solution algorithm to other similar problems, and his progress resulted, most probably, from his *mnemic* rather than *intellectual processes.*

Here follow several records, illustrating the process of patient B's training.

The patient is given the problem: *Three boys found 87 mushrooms. The first boy found 42 mushrooms, the second boy found 12 mushrooms less. How many mushrooms did the third boy find?*

All the attempts to repeat the statement ended in failure: "Three boys found... I don't know... I've forgotten. (The problem is read once more.) Three boys found 48... no, 87 mushrooms, the second boy found 42 mushrooms... and the third one 12 mushrooms less. How many mushrooms did they find together?" (The patient gave a positive answer to the question if the problem was repeated correctly.)

The final question reproduces the value given in the statement.

The patient is given the text of the problem. *Read the problem.* Having read the problem, the patient immediately tries to solve it. "Yes, the problem is simple, but I am all the time forgetting the numbers. So, 87 ÷ 12 = this is how many mushrooms the third boy found." *Why have you divided?* "I've divided all the mushrooms by the second boy because he found least of all. That'll make... six... no, not six... seven... no, it does not divide without a remainder, therefore, it's wrong. But it's interesting why?! Then let it be 42 ÷ 12 = 3... no, it does not divide without a remainder either."

The word "less" causes an inadequate fragmented operation of division.

The operations are disconnected with the semantic system of the problem.

Look at the statement of the problem attentively. Should you learn about the third boy in one try? Maybe something else is unknown? (The patient consults the statement.) " No, everything is known but for the third boy." *Tell me what is known in the problem.* "All the boys found 87 mushrooms. The first boy — 42 mushrooms, the second boy — 12 mushrooms less. Oh, yes... there is another unknown value here . . . then it should be this way . . . we should learn how many mushrooms the second boy found . . . 87 ÷ 12 equals . . . but it does not divide without a remainder." *And why are you dividing?* "And what should I do, should I add?... well, I'll try to add: 42 + 12 = 54... the second boy found 54 mushrooms. And now we'll learn how many mushrooms the third boy found — 54 mushrooms ÷ 12 = ... no, it does not divide without a remainder either." *You have been solving the problem incorrectly. Let us study again. Read the problem.* (The patient reads the problem.) *Write down the scheme of the problem.* (The patient writes down the scheme with the help of the therapist.)

Correct recital of the problem does not lead to the correct operations.

He repeats the previous operation. The patient is searching for the proper operations beyond the system of the problem's logical relationships. He inadequately manipulates with the operations of addition and division.

Three boys found together — 87 mushrooms
The first boy — 42 mushrooms
The second boy — 12 mushrooms less
The third boy — x mushrooms

What is known in the problem? "Everything is known, only the third boy is unknown." *Look at the statement attentively.* "Well, the second is also unknown." *What should you learn first?* "How many mushrooms the third boy found." *Can you learn this in one try?* "Yes, I can, it's easy: 87 ÷ 3, that's all."

The patient does not analyze the problem. He once again grapples with meaningless arithmetical operations.

He repeats the previous operation.

Do you know how many mushrooms the first boy found? "Yes, I do." *How many?* "42 mushrooms." *Do you know how many mushrooms the second boy found?* "No, I don't." *What is known about him?* "He found 12 mushrooms less." *Less than who?* "Than... than... the first." *Can you learn how many mushrooms the second boy found?* "Yes, I can... 42 mushrooms... plus..." *Why add?* "Well, then to subtract, 42 mushrooms − 12 mushrooms = 30. The second boy found 30 mushrooms." *Go further.* "What further? That's all." *No! Consult the statement.* "We should learn how many mushrooms found... but I've already said, 87 ÷ 3. "*How many mushrooms did they find together?* "87."

An inadequate operation.

How many did the first boy find? "42." *How many did the second boy find?* "30. Yes, 87 − 30." *No, look here.* (The investigator writes: Three boys = 1 + 2 + 3.) *Let us substitute 87 mushrooms = 42 + 30 + ?* "Well, that's it, 42 + 30 = 72 mushrooms 87 − 72 = 15. The third boy found 15 mushrooms."

Specific assistance rendered by the investigator results in the correct solution of the problem.

Is it now clear how to solve the problem? "Yes, it is." *Repeat its solution on your own.* (The patient fails to do this.)

While solving this problem again on his own, the patient persisted with the same mistakes. He managed to solve it correctly unaided only during the third session. But even then he, nevertheless, failed to work out the analogous problem: *Three girls caught together 55 butterflies. One of them caught ten butterflies, the second girl caught five more butterflies. How many butterflies did the third girl catch?*

When the patient was asked if he had solved a similar problem he gave a negative answer.

It took us five sessions to teach the patient to work out the problem: *One schoolboy bought one ABC book and one pen and paid 37 cents for them. Another one bought one ABC book and two pens and paid 49 cents for them. How much do one ABC book and one pen cost separately?*

At the first session, the patient managed to work it out correctly with the investigator's repeated assistance. But this is how he solved the same problem the next day:

Have you read the problem? What should you do next? "Write down the scheme." *Please, do.* (The patient writes down the statement of the problem correctly.) *What is unknown to you here?* "How

Recurrent operations irrelevant to the system of the problem.

much a pen costs and how much an ABC book costs." *Consult the statement. What accounts for the difference between the boys' purchases?* "It's a pen... the pen costs ten cents." *How did you learn this?* "I calculated mentally." *No, write it in your notebook, please.* "Okay, so...one pen costs 10 cents. 10 + 10 = 20...two pens cost 20 cents. And now the ABC book, it must cost 30 cents." *How did you find it?* "I think so." *Do you remember how we solved this problem yesterday?* "Yes, I do. I solved it the same way."

The deficiency in searching activity necessary to solve the problem.

The process of training is repeated:

- The statement is written down again.
- The purchases of the two boys are compared once again.
- The patient's attention is once again drawn to the difference in the purchases and the money spent.
- The investigator points to the essentials of the problem and even spells out the interrelationships between the components of the problem using real objects.

This procedure resulted in the correct solution of the problem with the investigator's assistance.

The training procedure was repeated for five days. As a result the patient correctly worked out this specific problem on his own at each subsequent session.

The investigator's attempts to generalize the solution method ended in failure. In addition to the correct solution of the above problem, the patient persisted with the following "solution" of a similar problem: *One girl bought one meter of ribbon and one meter of braid and paid 69 cents for her purchase. Another girl bought one meter of ribbon and two meters of braid and paid 75 cents for them. How much do one meter of ribbon and one meter of braid cost separately?*

"We should learn how much one meter of braid costs...75 cents − 69 cents = 6 cents. Aha...one meter... of braid — six cents and there are two meters of braid here. Therefore, 6 + 6 = 12 cents. So, one meter of ribbon costs 12 cents." *It's wrong.* "Is it? Well, then 6 cents + 6 cents = 12 meters of braid...but it is said here "two meters of ribbon." Therefore, two meters of ribbon cost 12 cents. 75 − 12 cents = 63, this is one meter of braid. That's all. Now it is correct." *Have we solved a similar problem?* "No, we have not." *Isn't it similar to the problem we have just worked out — with the ABC book and the pen?* "No, it is not." *Did you solve the problem*

The first operation is performed correctly.

The fragment of the problem worded as "two meters" led to the inadequate operation of addition disconnected with the general idea of the problem.

The patient can't adequately evaluate his solution.

correctly? "Yes, I did. Now it's correct. I've been turning it this way and that way, but I've solved it after all."

The above protocols graphically show that we had to train the patient in the specific solution method for each problem. With this in mind, we taught him to analyze a statement, to process the received information, to compare the data provided in the statement, and to draw the necessary conclusions. The direct instruction to perform certain operations with an emphasis on analyzing the statement was sufficient for patients with other forms of the frontal syndrome, but did not work in this case. But even this extended assistance rendered by the investigator did not result in the mastery of either *a general method* for solving problems or *a specific solution algorithm* which could be later transferred to similar problems. It turned out that this patient was unable to benefit from any training.

Hence, our experiments in rehabilitation training of patients with the frontal syndrome allowed us, not only to develop a better insight into the structure of their intellectual disorder, but also has contributed to the elaboration of methods for compensating for the impaired intellectual activity. Our research has shown that this group of patients suffered most from the disturbed organization of behavior on tasks involving goal-directed activity. For example, an arithmetic problem did not induce the required systemic activity aimed at the analysis of its statement. Such patients could be easily diverted by irrelevant associations, arising impulsively or by perseverative thinking. They encountered difficulty comparing their actions with their original intentions and were unable to correct their mistakes. This resulted, in an initial severely impaired orientation to the problem statement, which apparently disturbed their solution strategy and control over intellectual operations. The methods of rehabilitation training provided external programming of their general behavior during the course of problem solving and permitted compensation for their impairments.

During training, patients from both groups (those with the prevailing syndrome of disinhibition and impulsiveness, on the one hand, and those with the syndrome of inactivity, on the other) mastered general problem solving. These methods, however, did not become internalized. Therefore, the training of such patients did not lead to the development of self-regulating systems. The withdrawal of the support of an external program resulted in the disintegration of their intellectual activity.

The experiments in rehabilitation training of patients with a massive frontal syndrome, once again, confirmed the dramatic difference between the disturbed intellectual operations caused by the lesion of gnostic (parietal-occipital) sections of the brain, and the deep disorders in the structure of the complex purpose-oriented activity caused by an injury to the frontal lobes.

---

[1]The sign " + " stands for the correct repetition of the respective part of the statement.

Chapter 10

DISORDERS IN PROBLEM SOLVING IN LOCAL BRAIN LESIONS

Conclusions

We have described the disorders in problem solving caused by different localized focal brain lesions. For the purpose of analysis, we selected patients with localized brain lesions, on the one hand, within the posterior-occipital area of the left hemisphere, and on the other, in the frontal lobes. The latter suffered from different variants of the frontal syndrome. A comparative analysis of their problem solving impairments allowed us to draw some basic conclusions which are summarized below.

The process of problem solving, which is a typical intellectual activity, may become impaired following an injury to virtually any area of the brain, particularly when the injury involves its dominant (left) hemisphere. The characteristics of problem solving disorders (like that of any intellectual impairment) varies considerably. The nature of the impairment and the element in which the intellectual process is impaired is largely determined by the component of complex intellectual activity, which is disturbed as a result of a particular focal brain lesion.

The research findings of modern neuropsychology (see Luria, 1962; 1963) convincingly prove that different cerebral areas, whose joint functioning enables complex mental activity, have strictly specialized functions. Each of these areas makes its specific contribution in structuring this complex activity. Therefore, brain lesions localized to different brain areas can cause different impairments in complex functional systems.

Disorders in Problem Solving Following Lesions of the Parietal-Occipital Areas of the Brain

The parietal-occipital areas of the brain form the cerebral end of the visual-spatial analyzer. They enable the reception and processing of visual-spatial information and its synthesis into simultaneous (spatial) groups. Therefore, an injury to these sections of the brain results in the disintegration of our ability to integrate the elements of successively perceived information into simultaneously visible schemes. As a consequence, all the activities which need such integration are disturbed, while the mental processes which do not require such syntheses may remain intact.

Hence, patients with injured parietal-occipital areas of the brain enjoy stable intentions and purposeful behavior. They work hard on solving a problem and can critically compare the results of their activity with the original intention. Such patients are, however, unable to find their bearings and begin to encounter difficulty when integrating separate impressions into a single entity. They suffer from an inability to clearly tell right from the left, to read a geographical map, etc.

We have witnessed similar disorders in their speech and sometimes arithmetic skills; such patients are unable to immediately grasp the meaning of complex logico-grammatical structures as their understanding depends on simultaneously perceiving systems of words which form complex logico-grammatical relations. These patients lose insight into the place value of numbers and cannot do mental arithmetic. Whenever they come across such problems, which are easily solved by normal persons, they have to resort to a long detailed series of consecutive operations that compensate for their impaired comprehension of complex logico-grammatical and mathematical schemes.

The above defects are very pronounced during the *process of problem solving*. Patients with parietal-occipital lesions of the brain are quite willing to begin performing the given assignment. They attentively listen to the statement of a problem or read it; they do not forget, as a rule, the main question of the problem and are able to remember what is unknown, according to the statement, and what should be the goal of their activity.

Although they listen to the statement attentively, they fail to grasp the interrelationships inherent in it. Such patients cannot understand the logico-grammatical constructions of a statement and often fail to immediately grasp even such, it would seem, the elementary construction "one boy had so-many apples and his brother had so-many" (to say nothing of "so-many times" and "by so-many units"). Therefore, they must carry through extensive work on *decoding the statement of a problem* and can develop an insight into it only due to a complex series of successive operations performed with the help of a number of auxiliary aids. This work is time-consuming, it demands great persistence and often constitutes the main difficulty encountered by such patients in the process of problem solving.

If this stage is successful, patients with parietal-occipital lesions of the brain encounter no difficulty distinguishing the essential components of the statement, can easily chart the semantic scheme of problem solving, and perform the required arithmetical operations.

At this stage of intellectual activity, they may once again face certain difficulties caused by impairments in their arithmetical operations, but they generally overcome these difficulties, especially if they are taught to utilize algebraic methods of problem solving, which are not as fraught with difficulties related to the performance of specific calculations.

It is noteworthy that patients of this group generally retain a critical attitude to their performance, actively compare the obtained results with the original statement, and correct their mistakes. Their operations are always selective and determined by the main objective.

It is precisely for this reason that specifically designed programs, which facilitate the understanding of respective logico-grammatical relationships and the performance of respective operations, allow for successful compensation for their defects and provide the conditions for restoring the disturbed form of activity.

The nature of problem solving disorders caused by the frontal lesions is quite different.

Disorders in Problem Solving in Patients with Lesions of the Frontal Lobes

We'll begin the review of the specific problem solving disorders characteristic of patients with injured frontal lobes with analysis of the initial element of the problem solving process, that is, repetition of a statement.

The statement of a problem always has a typical structure: it comprises concrete data which makes up the "material" content of the statement and ends in a *question*. This question, which constitutes the *predicative element of a problem*, establishes certain relationships in the data of the statement, integrates them into a single whole, and makes the problem *meaningful*. When repeating a problem, normal people may forget some of the "material" components of its statement, but they never forget the main question. It is precisely for this reason that a normal subject always knows *what* the problem was given him *for*, *what* he is to *do* with the specific data of the statement, and *what* should be the *direction* of its strategic solution.

The same is true of patients with injured posterior (temporal-parietal-occipital) sections of the brain. They may forget certain details of a statement, they may encounter difficulty in recollecting its numeric or materialized components and reproducing their interrelationships, but they always preserve the relatively intact predicative structure of a problem and its final *question*.

We witness the opposite situation in the repetition of a statement in the case of patients with lesions of the frontal lobes. They are relatively good at retaining the essential elements of a statement, although they may usually simplify the relationships mentioned in it or alter them under the impact of previous inert stereotypes. But these patients generally face the *greatest difficulty in memorizing the question of a problem*. Therefore, the whole problem deprived of its semantic structure easily turns into information on data insubordinate to any question which this data should help to answer.

This specific phenomenon is related to disorders in the *predicative structure of speech* described in the review of aphasia caused by an injury to the anterior sections of the brain and disorders in intellectual dynamics typical of an injury to the cerebral areas located in anterior of the Broca area (see Luria, 1963, pp.296–339). It is quite probable that these disturbances take root in impaired *internal speech* and its disordered predicative structure which were thoroughly studied by Vygotsky (1934, 1956).

The impaired predicative structure of a problem, typical of the vast majority of people with injured frontal lobes, reveals itself at the stage of *repetition*.

When repeating the statement of a problem, many patients suffering from an injury to the frontal lobes either completely omit the question or replace it with the echolalia of one of the terms mentioned in the statement, or in the most severe cases, replace the logical question determined by the statement with some irrelevant question which has nothing to do with the context of the problem.

As shown by the earlier illustrations, we have repeatedly come across cases of patients repeating a complex statement then limiting himself to the reproduction of its constituent data and stopping (perfectly unaware of the incompleteness of such a reproduction). When asked, "What's then?" the patient would answer, "But that's all!" In the tests, we have more frequently come across cases when a subject, repeating a problem, substituted the echolalia of initial data for a meaningful question. Here's how one of our subjects repeated the problem: *Some collective farmers had a 12 hectares plot; they harvested three tons of grain from each hectare and sold one-third of the harvest to the state. How much grain was left?*

"Some collective farmers had 12 hectares of land; they harvested three tons of grain from each hectare and sold one-third of the harvest. How many hectares of land did they have?"

Such wording of the statement retains all the material elements of the problem, but makes it absolutely meaningless.

In our view, the disintegration of the predicative (semantic) structure of a problem is most pronounced in the special experiments when the subjects are requested to *make problems on their own*, modeling them on the previously elaborated ones.

Patients with injured posterior (temporal-parietal-occipital) areas may make blunders in reproducing the material data of a problem. They may encounter difficulty in naming the respective objects or numbers or memorizing the necessary correlations, but such people, as a rule, never forget the predicative wording of the question. Whereas patients with injured frontal lobes, on the contrary, may recite the "material" part of the statement well, but ask no question at all. As a result the structure developed by such subjects ceases to be a problem.

Similar data may be obtained in experiments when a patient is given only a statement and requested to *formulate, on his own, the ensuing question*. These tests reveal the impaired predicative structure of a problem typical of subjects with frontal lesions of the brain: many of them either do not formulate any question at all, or reproduce in the form of the question any initial data mentioned in the statement, or say that it is possible to put any question to this problem. Here follow several illustrations of such impairment.

Patient U is given the statement: *There were 150 vases in a shop; some of them were sold and 65 vases were left.*

Then he is asked: *What should we learn?* "How many vases were there in the shop . . . " *Is it really so? Think, please!* "How many vases were left in the shop. . ."

The same patient is read the following statement: *A shop sold 750 kg of apples, thereafter, there were four boxes of apples with 20 kg left in each.*

He is again asked: *What should we learn?* The patient answers: "We may learn a lot here. . . for instance, at what price did they sell their apples? Or, maybe, how many apples remained unsold? It's up to us—we may learn whatever we want."

In all the above cases, the wording of the question is altogether lost, and the statement is deprived of its predicative structure.

In other cases, the question of a problem is formulated, but it is replaced with *a different question* undetermined by the overall semantic structure of a given problem.

To this type of semantic alteration of the question, we may refer to numerous above cases when the question to the problem was replaced with the perseveration of the question to the previously solved problem, or when the final question was replaced with an intermediate specific question directly ensuing from the text of the statement.

The former alteration may be illustrated by the case of a patient with a massive lesion of the frontal lobes. He first worked out the problem: *A boy had 12 kg of apples and he gave one-third of them away. How many apples were left?* After he had solved this problem he was given another problem to solve: *A housewife uses six liters of milk in five days. How much does she use in a month?* Here's how he repeated this problem: "A housewife uses six liters of milk in five days. How many liters were left?"

The latter alteration may be illustrated by the numerous cases when the question of this problem: *A son is five years old. In 15 years his father will be three times older than the son. How old is the father now?* is perceived as "How old will be the son if he is three times older?" or "How old will the son be in 15 years?"

The former variant of the altered repetition of a problem shows that the overall semantic system of a given problem is not analyzed at all; it is replaced with an inert reproduction of the previous problem's stereotype. We have never come across such an impairment in the case of patients with injured posterior (temporal-parietal-occipital) sections of the brain. This fact testifies to the inactivity and pathological inertness of established stereotypes which are typical of the frontal syndrome.

The latter variant of the alteration points to an incomplete analysis of a statement and the ease with which the solution algorithm is simplified, the data of the statement is disregarded midway, and the final question of a problem is replaced with some intermediate questions directly determined by certain statement elements. Such mistakes in the wording of the question may be found in all sorts of brain lesions. However, the point is that patients with temporal-parietal-occipital injuries easily correct mistakes after the second reading of the problem, while people with the frontal syndrome and inactive thinking, suffer from inert stereotypes and encounter serious difficulty in correcting such mistakes.

The above difficulties in the repetition of a problem's statement and in preserving its predicative (semantic) structure bring us to the second feature which is more or less characteristic of patients with lesions of the frontal lobes. This is the *defective preliminary analysis of the statement and the loss* or *instability of the orientation basis for an intellectual operation*. We have already mentioned that a normal subject, similar to a patient with temporal-parietal-occipital lesions (see Chapter 2), always precedes his problem solving by the careful analysis of a statement. Patients with injured posterior areas often encounter serious difficulty in understanding the statement, but nevertheless, they never begin to solve a problem without such preliminary analysis and carry out extensive (and usually highly emotional) work on the analysis of a statement.

The patients with injured frontal lobes of the brain behave quite differently. If the statement of a problem directly and explicitly determines its solution algorithm (which is, for instance, the case of simple problems) such people solve it at once. Whereas if the problem does not determine such an explicit solution algorithm and requires the preliminary analysis of its statement, the patients of this group generally do not even try to orient themselves in the statement, distinguish its essential components, and identify the basic interrelationships. Instead, they distinguish some directly perceivable fragment of the statement and begin to perform operations dictated by this fragment, substituting such a "solution" for the orientation basis of an operation and the search for a proper solution algorithm. If the investigator points to the incorrectness of this approach and the need for the preliminary analysis of a statement, such patients do not change their behavior, but simply turn to a different fragment of the problem (which then turns into the initial point for their impulsively arising attempts at the direct "solution" of the problem). The loss or instability of the orientation basis of an intellectual operation may assume different forms and may be more or less apparent depending on the variant of the frontal syndrome. But one of the most fundamental findings of our study is that this is precisely the element that suffers most in lesions of the frontal lobes.

The impaired orientation basis for an operation is closely related to the *third* specific feature apparent in most frontal lesions. Such people suffer from an *inability to draw up the solution plan*. This inability was observed virtually in all the cases of pronounced frontal syndrome.

One of the authors (Tsvetkova, 1966) studied how patients with local brain lesions understand literary texts and make stories on their own. She showed that the two groups of patients above drastically differed in this type of activity as well. Whereas patients with temporal-parietal-occipital lesions can easily draw up the plan of the respective story, but for certain reasons encounter difficulty in its coherent retelling, patients with injured frontal lobes generally fail not only to make a more or less long story on their own, but also to draw up its plan. Her research findings prove that the drawing up of the story plan (program) is one of the most difficult forms of activity for such patients. Even if the investigator persistently urges them to phrase such a plan, they do their best to bypass it and go directly to retelling it proper, since they are absolutely unable to formulate the preliminary retelling plan.

We have noticed patterns in our experiments in problem solving. If the investigator asks a patient with injured posterior (temporal-parietal-occipital) sections of the brain to draw up a solution plan for a more or less complex problem, the patient begins to analyze its statement and tries (although often unsuccessfully) to determine and formulate its solution algorithm. Anyway, such a patient never begins to solve a problem without determining its general solution scheme.

In contrast, a patient with the pronounced frontal syndrome behaves quite differently. When listening to the request to draw up the solution plan for a complex problem, such a person generally begins to solve it at once, making abortive fragmentary attempts at its solution. The repeated urge to formulate the solution plan of a problem, as a rule, results in the repetition of separate statement elements or the wording of the problem's questions. Furthermore, such a patient often alters these questions and replaces them with the initial data mentioned in the statement. That is why patients with injured frontal lobes can answer the questions of the plan offered to them in the ready form and solve the problems when the solution algorithm is spelled out to them, but they are unable to independently begin to analyze the statement and draw up its solution plan.

This intellectual peculiarity of people with frontal lesions leads us to the *fourth* specific feature of their intellectual processes: *disorders in the systematic operations of problem solving hierarchically subordinated to the respective solution program.*

As noted above, intellectual activity may be defined as an organized process of problem solving, which relies on a system of interconnected operations. This is precisely the process that suffers in the case of a frontal lesion.

Therefore, here lies the drastic difference between the two groups of patients under review. Patients with injured posterior (temporal-parietal-occipital) sections of the brain generally proceed in their problem solving from an earlier developed plan. They distinguish the general solution scheme and although such people exhibit serious difficulties in performing the consecutive series of interconnected operations (caused either by the impaired simultaneous or successive syntheses or unstable mnemic processes or speech deficits), their operations generally *never disagree with the basic program*, are *always determined by it,* and are *selective in character.* Such people, as a rule, notice and promptly correct infrequent deviations from the planned solution of the problem.

This contrasts sharply with the picture when we review the problem solving of patients suffering from an injury to the frontal lobes. Since they do not rely on a stable program of operations based on the analysis of the statement, patients with an apparent frontal syndrome are easily affected by factors which are irrelevant to the solution algorithm.

In some cases, such patients distinguish from the statement directly perceivable fragments and try to "solve" the problem with the help of disconnected fragmentary operations.

In other cases, they are affected by the inert stereotypes developed while solving previous problems or performing some previous operation of the given problem. Therefore, they substitute the inert reproduction of these stereotypes for a true solution of the problem.

In still other cases, they abandon all attempts to determine the solution plan and replace problem solving with impulsive guessing or separate numerical operations disconnected with the statement, which result in extracontextual attempts to perform certain operations on the numbers mentioned in the statement.

The specific feature of such attempts is that patients may completely disregard the denominations of numbers which determine their place in the semantic system of a problem, and for instance, add hectares to kilograms, subtract weight from the number of objects, and perform any actions apparently disconnected with the original statement.

In the most pronounced cases of the frontal syndrome, this disintegration of the programmed nature of operations is aggravated by the occurrence of *irrelevant operations* undetermined by the statement. In these cases operations performed by the patient are completely deprived of their selective character, and we witness the discontinuation of any attempts at an organized intellectual process.

One of the most typical peculiarities of intellectual disorders in frontal lesions is the impaired programming of activity and the loss of the selective nature of its constituent operations.

The last distinguishing feature typical (to some extent) of all the patients with the conspicuous frontal syndrome concerns their *defective understanding of the nature of the performed operation and their inability to independently correct the mistakes made.* As shown above, people with injured temporal-parietal-occipital sections of the brain remain perfectly aware of the nature of performed operations while working on a task (including problem solving). These patients determine the plan of respective operations, continuously compare the performed operations with the original plan, and correct their mistakes. It is precisely this structure of activity that characterizes rehabilitation work with such patients and lays the basis for its success.

The structure of activity typical of patients with the pronounced frontal syndrome is quite different. These people, as a rule, neither pay attention to the analysis of the statement nor determine a sufficiently clear solution plan. They are generally unable to conscientiously analyze the performed operations or to formulate them. Even when such patients practically solve the given problem, as a rule, they are unable to tell by what methods they have obtained the desired effect and simply repeat instead the performed operations.

This impairment is most conspicuous in the analysis of the obtained results and the correction of the mistakes made. *The process of comparing the obtained result with the original statement of a problem is generally the weakest link in the structure of intellectual activity of patients with the frontal syndrome.* This operation, which requires stable feedback connections and an intact system of the acceptor of the operation, most frequently falls out altogether; in many cases it is fragmentary in character and may be easily replaced with the statement of immediately apparent differences.

Two of the most typical peculiarities of the mental disturbances caused by frontal lesions, which have long been observed in clinical work as "a disordered critical attitude," are the impaired assessment of the operation performed and the mistakes made which lead to an inability to correct them on one's own.

Whereas the former defects may be reduced to the *defective programming* of intellectual activity, the latter concerns the *defective regulation and control* typical of such patients.

It is obvious that the above data completely agrees with the general characteristics of the mental disorders caused by the frontal syndrome (see Luria, 1962, 1963; Luria and Kholmskaya, 1966).

We are still unaware of the physiological mechanism which underlies the impairments described, but we are sure that this characteristic agrees with the clinical facts and that the above impairments in intellectual processes caused by lesions of the frontal lobes supplement it with significant facts.

Disorders in Problem Solving in Different Variants of the Frontal Syndrome

We have characterized, in general terms, the changes in the intellectual activity of patients with injured frontal lobes manifested in their problem solving.

Now let us summarize the differences in the actualization of intellectual processes caused by different variants of the frontal syndrome. Their evaluation will allow us to take a major step toward distinguishing the factors underlying the above disorders.

It is well known from clinical experience that focal brain injuries may be manifested differently depending on their nature and the pathophysiological background against which they are observed. Clinicians are equally aware of the dramatic difference in the manifestations of a mental disorder depending on whether it occurs against a background of *irritation* or *inhibited conditions,* which may reveal themselves in adynamia or the increased inertness of nervous processes.

These pathophysiological peculiarities may have a tangible impact on the development of syndromes caused by different frontal lesions. It is, however, important to note that differently localized lesion focuses usually have different effects. They often lead to changes in higher nervous processes characterized by the domination of one of the above pathological states.

It is precisely for this reason that, fully aware of the relative nature of the symptoms described, we can speak about the structurally different variants of the frontal syndrome caused by the topical heterogeneity of these injuries.

We have mentioned above a great variety of disorders caused by an injury to the basal, posterior-frontal, and prefrontal cerebral areas. We have also discussed the severe aggravation of the frontal syndrome if a lesion caused by the main focus is accompanied by general cerebral impairments (for example, hypertension-dislocation manifestations). Now we should summarize the specific features of intellectual processes observed in these cases.

An injury to the *basal (or medial-basal) sections of the frontal area* results, of course, in conspicuous changes in the patient's emotional sphere and behavior, but does not really affect the actualization of his intellectual processes. Therefore, in the least severe forms of this group of lesions (tumors of the olfactory fossa, arachnoendothelioma of the basal sections of the frontal area, and the slowly growing intercerebral tumors — astrocytomas or oligodendrogliomas) intellectual disorders may be either inconspicuous or very indistinct. They are much more pronounced in more severe cases.

The most apparent defects of intellectual activity manifest themselves in impulsive intellectual operations and the impulsive drawing of separate conclusions unsubstantiated by adequate analysis of a statement. They may reveal themselves when the solution algorithm of a problem is not explicity determined by its statement, and when it is necessary first to decode the directly perceived statement in order to get the desired answer.

We have already described in another publication (see Luria, 1963, p.344) the changes which occur in such patients' intellectual operations when they receive assignments: the overt wording disagrees with the required solution algorithm.

As noted above (Chapter 3), such patients generally understand the statement well. Only when working out the most difficult problems may they (as a result of the unstable preliminary stage of statement analysis) simplify the solution plan and make preliminary attempts at problem solving which generally assume the form of disconnected specific (fragmented) operations that are insufficiently coordinated with the statement as a whole. It is characteristic of these cases that even minor assistance, often just an additional question whether the patient has solved the problem correctly, is enough to make him compare the operation performed with the original statement and correct his mistake. Such patients fail to solve only the problems in which the determination of the solution algorithm is complicated by certain factors.

Our observations have proven that such patients feature neither pathological inertness of established stereotypes nor loss of the selectivity of operations performed. They may be caused only by an aggravation of local brain lesions by additional general cerebral factors (intoxication or hypertension-dislocation).

It is relatively easy to compensate for intellectual impairments these patients by encouraging the analysis of a problem and providing conditions for problem solving which strengthen the orientation basis of an operation and inhibit their immediate impulsive answers (for instance, to ask them to solve a problem in writing or to program their successive behavior).

The above impairments differ considerably from intellectual disorders caused by lesions of the *posterior-frontal sections of the brain* (and, above all, the posterior-frontal sections of the dominant hemisphere).

The effect of their most limited forms may be confined to the more volitional nature of intellectual processes and the increased inertness of established stereotypes which complicate the normal actualization of intellectual activity. In these cases, the operations of problem solving may assume a voluntary extended character and any attempts to contract an operation or transfer it to the automatized level, result either in considerable difficulties or, occasionally, in the substitution of the inert stereotype of the previously developed solution methods for the desired solution algorithm. The respective facts have been described in detail in another publication (see Luria, 1963, pp. 184–244; 340–380).

It is worthy to note that whenever lesions of the left posterior frontal areas are massive enough, all the above phenomena are most apparent in speech. The pathological inertness of established stereotypes reveals itself most graphically in verbal processes. The patient begins to alter both the statement as a whole and its separate components (numbers, denominations, and operations) by inert persevering stereotypes. Therefore, it is much more difficult for the patient to solve the problem although he preserves its general solution algorithm. It is thought provoking that the verbal solution of problems or their solution supported by loud speech are sometimes fraught with greater difficulty than mental arithmetic or an extended written solution in which verbal perseverations may be eliminated easier.

Other severe changes in the structure of intellectual activity are caused by lesions of the posterior-frontal areas of the left hemisphere which occur against a background of the patient's *general inactivity* and *the increased inertness* of established stereotypes (these phenomena are often intimately interconnected). In these cases, the above intellectual disorders are aggravated by an inability to independently begin the discourse, a tendency to stick to developed stereotypes, an inability to proceed from a performed operation to the next one, and an immediate refusal to carry out intellectual activity.

All these factors are also quite conspicuous in the process of problem solving where the desired algorithm is replaced with the inert reproduction of established stereotypes, and it is very difficult for the patient to move not only from a solution element to the following one, but also from one type of problem to a different one.

It is, however, characteristic that in most cases the posterior-frontal syndrome does not lead to any stable disturbances in the selectivity of intellectual processes. Generally, such patients keep in mind the original statement of a problem and suffer from the pathological inertness primarily due to the *operational elements* of intellectual processes. They continuously attempt to correlate them with the operations performed and, generally, preserve a sufficiently critical attitude to the obtained results. Such people very seldom overlook their mistakes, although they may be unable to correct them on their own. But they easily correct their mistakes provided additional encouragement or minor assistance.

It is precisely for this reason that in the case of such lesions there is always a chance to compensate for the defect described and to use the methods of programmed education which split the process of problem solving into a series of consecutive questions, and then to structure the patient's behavior by a detailed program which points to the successive elements of the respective operations.

We have already illustrated such intellectual impairments caused by an injury to the posterior-frontal sections of the brain.

This picture drastically differs from the clinical picture of intellectual disturbances caused by *massive lesions of the anterior sections of the frontal lobe.*

We must acknowledge that an injury to the prefrontal brain areas causes a wide variety of consequences. In this case, we witness intellectual disorders ranging from inconspicuous changes, brought about by slowly growing and relatively limited benign tumors which do not destroy the brain matter, to a severe disintegration of intellectual activity accompanied by the complete loss of the selectivity of intellectual processes. The latter may be most frequently caused by the bilateral lesion of the frontal lobes or rapidly growing intracerebral tumors when local brain lesions occur against a background of general cerebral phenomena (intoxication and hypertension-dislocation). Severe impairments of intellectual activity may also be caused by a bilateral resection of the frontal lobes or the growing atrophic process which leads to the impaired functioning of the frontal lobes. The cases of severe intellectual disorders are caused precisely by lesions of the frontal lobes (see Kleist, 1934; Brickner, 1936; Rylander, 1939).

We have discussed above in sufficient detail changes in intellectual processes caused by the continued growth of a tumor which spread from the right frontal lobe to the left frontal lobe and necessitated the resection, first of the right, and then partially the left frontal lobe. We have also analyzed in detail intellectual disorders caused by the massive atrophy of the white matter of the left frontal lobe.

Our research findings (confirmed by the study of cases with massive lesions of the frontal lobes) testify to a severe disintegration of intellectual activity, described in the first section of this final chapter.

It is worthy to note that intellectual disorders typical in this group of patients reveal themselves most graphically in the disintegration of their intellectual operations. For example, having listened to a problem, such patients generally do not analyze its statement and either immediately respond to some of its fragments or replace active analysis of the statement with the reproduction of an inert stereotype developed during the solution of previous problems.

Patients whose massive frontal syndrome occurs against a background of general aspontaneity and adynamia are unable to solve arithmetic problems. Their attempts at problem solving are limited either to the echolalia of the statement or the reproduction of inert stereotypes. Whereas if this syndrome occurs against a background of disinhibition, consistent problem solving is replaced with irrelevant associations or impulsive fragmented operations. In both cases, the operations substituting for consistent problem solving are performed without due comparison with the statement, and are no longer selective and purposeful in character.

Intellectual impairments caused by massive lesions of the frontal lobes primarily result in the disintegration of systemically interconnected operations against a background of both the disordered preliminary analysis of a statement and the impaired operations of comparing the obtained results with the original statement (in other words, *in the disordered programming and control of intellectual processes*).

The present study has highlighted only some of the problems related to impairments in intellectual processes in the different variants of the frontal syndrome.

We have neither touched upon the issues of specific intellectual changes caused by lesions of the *right* frontal lobe nor discussed in sufficient detail the problem of disordered intellectual operations caused by a focal lesion located deep in the frontal-diencephalic area. It is known that such patients suffer from massive changes in their consciousness, disorientation, and general disintegration of the selective system of connections, which brings their psychic processes down to a dream-like level of consciousness.

All these intellectual disorders will be specifically studied in our future publications.

Some Problems and Future Prospects

The study of problem solving disorders in local brain lesions has allowed us to describe different types of changes occurring in this complex form of intellectual activity, and has brought us closer to the psychological analysis of difficulties that may arise in the course of its performance. Thereby, neuropsychological analysis has proven its effectiveness, not only for clinical work (where it is widely applied to improve the accuracy of the symptomatic diagnostics of local brain lesions), but also for psychological research on the structure of intellectual processes.

The neuropsychological method for analyzing the structure of intellectual activity is in its infancy, and therefore, it is only logical that in this work we could not touch upon numerous problems of general psychology and pedagogy. Hence, we would like to simply chart out in the most general terms some of these issues and the prospects for their future study.

Throughout our research, we have seen that the *question* of the problem is of major importance for the further selective actualization of intellectual activity. This aspect of the problem merits a special study. We certainly should not limit ourselves to the description of the question and repetition in various forms of intellectual impairments, but analyze how it is understood by the subject and how he can *put respective questions to the problem on his own*.

Equally important is careful analysis of the *wording of a statement* which often plays the decisive role in problem solving as it can either simplify or complicate the understanding of a problem. We should also analyze in detail *methods* which can facilitate the understanding of the logico-grammatical structure of a statement.

The key issue is, however, the *psychological analysis of the problem structure* and the development of a *scientifically grounded classification of arithmetic problems* in terms of their psychological difficulty.

Although problem solving has existed as a school subject for several centuries, and the methods of problem solving have formed a specific discipline, the psychological analysis of different problem types, and the development of their scientifically grounded classification are still in their infancy. Most recommended methologies use the most primitive and inadequate techniques to study the given problems in terms of their difficulty. We are presently unable to name even a single satisfactory attempt to give teachers a scientifically based classification of the basic types of arithmetic problems which require operations differing in their psychological structure and are fraught with psychologically different difficulties.

Therefore, in the present study we had to limit ourselves to a relatively imperfect attempt to develop a set of problems differing in terms of their psychological difficulty.

It goes without saying that future studies must provide a rational, psychologically substantiated classification of arithmetic problems, and the methods of neuropsychological analysis will, of course, greatly contribute to its further elaboration.

We believe that this material of neuropsychological research will help scholars outline some promising lines of differentiated analysis of difficulties encountered by schoolchildren in the course of problem solving and develop some basic techniques for their elimination.

Every teacher, undoubtedly, knows that his pupils may face occasional difficulty in memorizing the question of a problem, decoding the logico-grammatical relationships of its statement, inhibiting impulsive guessing, or inadequate use of earlier developed stereotyped solution methods. The same difficulties were observed throughout our study, and we believe that their elimination is a priority objective of the academic process. It is equally true that the above materials make it self-evident that it is important to develop in pupils the activity of control and self-control, impaired in the case of intellectual disorders. In the future, we should work specifically to elaborate the basic methods for its development.

In this publication, devoted primarily to the psychological analysis of the process of problem solving and its constituent factors, we could merely outline the basic principles underlying the *programmed restoration* of this process, which is otherwise impaired in the case of different brain lesions.

We hope that this study will also be useful for future work along these lines and will make some contribution to the psychological analysis of programmed training in problem solving.

REFERENCES

1. Ajuriaguerra J. & Hecaen H., La cortex cerebral. Paris, 1960.
2. Anokhin P.K. (1949), Problems of higher nervous activity. Moscow (in Russian).
3. Anokhin P.K. (1955), New data on the specifics of the afferent apparatus of the conditioned reflex. Voprosi Psikhologii, N.6.
4. Bernstein N.A. (1957), Some appearing problems in the regulation of the motor acts. Voprosi Psikhologii, N.6
5. Bekhterev V.M., (1905–1907), The fundamentals of the teaching on brain functions. Issues I–VII, St. Petersburg (in Russian).
6. Brickner R.M., The intellectual functions of the frontal lobes. New York, 1936.
7. Bubnova V.K. (1946), Impaired understanding of grammatical structures in brain lesions and its restoration in the course of rehabilitation training. Moscow (in Russian).
8. Fulton J.F., A note on the definition of the "motor" and "premotor" areas. Brain, vol.52, 1935.
9. Fulton J.F., Physiology of the nervous system. Oxford, 1943.
10. Galperin P. Ya. (1957), An intellectual action as a basis for thought and image. Voprosi Psikhologii, N.6.
11. Galperin P.Ya. (1959), Continued research on the formation of intellectual actions in: Psychological Science in the USSR. Vol.1, Moscow.
12. Goldstein K., Aftereffects of brain injuries in war, New York, 1942.
13. Halstead W.C., Brain and intelligence. Chicago, 1947.
14. Head H., Aphasia and kindred disorders of speech. Vol. 1–2, Cambridge, 1926.
15. Leontiev A.N. & Galperin P.Ya (1964), The theory of knowledge assimilation and programmed education. Sovetskaya Pedagogika, N. 10.
16. Kleist K., Gehirnpathologie, Leipzig, 1934.
17. Luria A.R. (1946), On the pathology of grammatical operations. Izvestiya APN RSFSR, issue 3.
18. Luria A.R. (1946), Pathology of arithmetical operations. Izvestiya APN RSFSR, issue 3.
19. Luria A.R. (1947), Traumatic aphasia. Moscow (Russian; English translation, Mouton, The Hague, 1970).
20. Luria A.R. (1948), The restoration of brain functions after a war injury. Moscow (in Russian).
21. Luria A.R. (1950), The essays on the psychophysiology of writing. Moscow (in Russian).
22. Luria A.R. (1959), Development of speech and formation of psychological processes. In: Psychological Science in the USSR, vol.1, Moscow.
23. Luria A.R. (1960), The impairment of writing and calculation in case of inert higher processes (unpublished).

24. Luria A.R. (1962), Higher cortical functions in man. Moscow (Russian; English Translation, by B.Haigh, Basic Books and Plenum Press, New York, 1966).
25. Luria A.R. (1962), The disturbance of problem solving processes in massive lobal lesions. Dokladi APN RSFSR, N.5.
26. Luria A.R. (1963), The human brain and mental processes. Vol.1, Moscow (in Russian).
27. Luria A.R. & Kholmskaya E.D. (1962), The impairment of intellectual operations in the lesions of posterior-lobal area. Transactions of the APN RSFSR, N.6.
28. Luria A.R. & Kholmskaya E.D. (editors), Collected papers "The lobes and the regulation of psychic processes" (in Russian).
29. Maysel I.I. (1959), Intellectual disturbances in case of disintegrated active purposeful activity. Master Thesis (unpublished, in Russian).
30. Pavlov I.P. (1951), Lectures on the functioning of bigger hemispheres. Collected works, vol.4, Moscow-Leningrad (in Russian).
31. Potzl O. Die Aphasielhre Optisch - Agnostische Storungen, Leipzig, 1928.
32. Pribram K.H., On the neurology of thinking. Behavioral Science, vol.4,1958.
33. Rubinstein S.L. (1958), On thinking and methods of its research. Moscow (in Russian).
34. Rylander G., Personality changes after operations on the frontal lobes. London, 1939.
35. Sechenov I.M. (1878), The elements of thought. In: Vestnik Evropi.
36. Shemyakin F.N. (1954), Research on the topographic notions. Izvestiya APN RSFSR, issue 53.
37. Shemyakin F.N. (1959), Spatial orientation. In: Psychological Science in the USSR, vol.1, Moscow (in Russian).
38. Shkolnik-Yarros E.G. (1945), Movement disturbances in the lesions of the premotor area, Ph.D. Thesis, Moscow (in Russian).
39. Tsvetkova L.S. (1965), Disturbances of the problem solving process in the lesions of parietal-occipital brain areas. Noviye Isslyedovaniya v Pedagogicheskikh Naukakh, issue III.
40. Tsvetkova L.S. (1960), Difficulties faced by patients with injured lobal and parietal-occipital brain areas in working with literary texts. In: The lobes and the regulation of psychic processes. Moscow (in Russian).
41. Vygotsky L.S. (1934), Thought and language. Moscow (Russian; English translation New York–London, 1962).
42. Vygotsky L.S. (1960), The development of the higher mental functions. Moscow (Russian, English translation by Progress Publishing House.) In: Psychological Science in the USSR, vol.1, Moscow, 1966.
43. Zeigarnik B.V. (1961), Pathology of thinking. Moscow (in Russian).

Note to the Reader:
Some of the information is missing from this reference list. If you have questions about locating any of the material cited, please contact the publisher.

ADDITIONAL SOURCES

Annaniev, Psychology of sensual cognition, Moscow, the Publishing House of the APN RSFSR, 1960 (in Russian).

Asratyan, The role of the cortex of the bigger brain in the phenomena of organic compensation, Moscow, Transactions of Moscow University, issue 3, 1947 (in Russian).

Bane, Aphasia and Methods for its elimination, Moscow, Medicine publishing house, 1964 (in Russian).

Benson, D.F., Geschwind, N., The alexias in: P.J. Vinken and G.W. Bruyn, Handbook of clinical neurology, v. 4, Amsterdam, North-Holland Publishing Company, 1969.

Bethe, A., Plastizitat und Zentren Lehre. Bethes Hand. d. norm. und pathol. Physiol., XVI, 1931.

Bogorodsky, V., The general course of Russian grammar, Moscow, 1935 (in Russian).

Cannon, W.B., The story of the development of our ideas of chemical mediation of nerve impulses, American Journal of Medicine, vol. 13, 1934.

Dale, H.H., Transmission of nervous effects by acetylcholine. (Harwey Lecture), Bull. N.Y. Acad. Med., 13, 1937.

Elkonen, D.B., Children's psychology, Moscow, Prosveshcheniya Publishing House, 1960 (in Russian).

Floyrence, P.J.M., Examen de phrenologie, Paris, 1842.

Goodglass, H., Studies in the grammar of aphasiacs, in: Development of applied psycholinguistic research, Rosenberg and J.H. Kaplan (eds), 1968.

Hebb, D.O., Man's frontal lobes, Arch. Neurol. Psychiatry, 54, 1945.

Hecean, H., Introduction a la neuropsycholoie, Paris, 1972.

Hershen, S.E., Klinische und anatomische Baitrage sur Pathologie d. Gehirns, Stockholm, 1922.

Jackson, H., Evolution and dissolution of the nervous system. Groonin Lecture, Selected Papers, 1884.

Kalita, N.G., Tsvetkova, Impairment and restoration of naming in acoustico-mnestic aphasia, Defectologia, N. 2, 1976.

Kertesz, A., Recovery patterns and prognosis in aphasia, Stroke, v. 9, N. 2, 1978.

Kock, E.P., Visual agnosias, 1967, Leningrad, Medicine Publishing House (in Russian).

Lashley, K.S., Brain mechanisms and intelligence, Chicago University Press, 1929.

Lebedinsky, M.S., Aphasia, apraxia, agnosia, Kharkov, 1941 (in Russian).

Leontiev, A.N., On the social nature of man's psyche, Vaprosi filosofii, N. 1, 1961.

Leontiev, A.N., Psychology of imagery, Transactions of Moscow University, issue 14, N. 2, 1979.

Lomov, B.F., The correlation between the social and the biological as a methodological problem of psychology, Vaprosi filosofii, N. 4, 1976.

Luria, Fundamental problems of neurolinguistics, Moscow, Moscow University Press, 1975 (in Russian).
Luria, Higher Cortical Functions, Moscow University Press, 1969 (in Russian).
Marie, P., Revision de al question de l'aphasie, Semaine Medicals, 1906.
Monakov, K., Uber Lokalization der Hirnfunktionen, Wiesbaden, 1910.
Morozova, N.G., On the understanding of the text, Izvestiya APN RSFSR, issue 7, 1947.
Nachamsohn, D., Cholintherase dans le system nerveux centrale, Bull. Sec. Chemical Biology, 21, Paris, 1939.
Paivo, P., On the functional significance of imagery, Psych. Bul., 73, 1970.
Pavlov, I., Twenty years of experience in the objective study of the higher nervous activity, Moscow, Medicine Publishing House, 1951 (in Russian).
Pavlov, Pavlov's Wednesdays, v.1-3, Moscow, The publishing house of the USSR Academy of Sciences, 1949 (in Russian).
Perelman, L.B., Physiological ways of rectification in traumatic disorder, Transactions of Moscow University, issue 3, 1947.
Seron, X., Aphasie et neuropsycholoie, Brussels, 1979.
Shekhter, Visual recognition, Moscow, Pedagogika Publishing House, 1981 (in Russian).
Sokolov, A.N., Inner speech and thinking, Moscow, Prosveshcheniye Publishing House, 1967 (in Russian).
Teplov, V.M., Psychology of aptitude for music, Moscow, the Publishing House of the APN RSFSR, 1957 (in Russian).
Tkachev, R.A., On the classification and prediction of aphasic disorders, in: Issues of clinical and patophysiological aphasias, Moscow, Medicine Publishing House, 1961 (in Russian).
Tsvetkova, Akhmetova, On the issue of comprehension disorders in motor aphasia, in: Psychological research in the practice of medical and labor examinations, Transactions of TSIETIN, Moscow, the publishing company of the Ministry of Health of the RSFSR, 1972 (in Russian).
Tsvetkova, Rehabilitation Training in Local Brain Lesions, Moscow, Pedagogika Publishing House, 1972 (in Russian).
Tsvetkova, Streltsyna, Speech comprehension disorders in aphasia, Defectologia, N. 5, 1981.
Vinogradov, V.V., Russian Language, Higher School Publishing House, Moscow, 1972 (in Russian).
Vvedensky, N.E., Collected Works, v. IV, book 2, 1938.
Wernicke, C., Der aphasische Symptomenkomplex, Breslau, 1974.
Wernicke, Lichtheim, Uber Aphasie, Arch. Klin. Med., v. 36, 1885.
Yakobson, Luria, in: Yakobson R. Selected Writings, Moscow, Progress Publishers, 1985 or Jakobson, R. Selected Writings, v. 2, The Hague - Paris, pp. 229-238, 1971.
Zhinkin, N. Mechanisms of Speech, Moscow, the Publishing House of the APN RSFSR, 1958 (in Russian).